ARTIFICIAL
INTELLIGENCE
AND THE FUTURE OF US

IMAGINATION NOT KNOWLEDGE IS KING

FRANCES MAHAN

ARTIFICIAL INTELLIGENCE AND THE FUTURE OF US
IMAGINATION NOT KNOWLEDGE IS KING

iUniverse books may be ordered through booksellers or by contacting:

iUniverse
1663 Liberty Drive
Bloomington, IN 47403
www.iuniverse.com
1-800-Authors (1-800-288-4677)

ISBN: 978-1-5320-8836-0 (sc)
ISBN: 978-1-5320-8835-3 (e)

Library of Congress Control Number: 2019918590

Print information available on the last page.

iUniverse rev. date: 11/26/2019

CONTENTS

ABOUT THE AUTHOR

FRANCES MAHAN WAS born in the Dominican Republic. She came to the US at an early age of 16th. Her family of three brothers and two sisters reside in Miami where she grew up.

As a young girl, Frances loved poetry, theater, singing and dancing, but never had she imagine to have or developed any interest in science and technology. Frances not only fail math in high school but she did not graduate from her paralegal program because she failed her accounting exam. Science was never one of her favorite subjects in high school or college. But little did she know, that in her future, she was destine to embark into a surprise of her own making, with her mind.

While residing in Dubai, The UAE with her husband of thirteen years; she discovered an insatiable need to learn about science from the local TV shows on the Discovery channel. It was here where she sat in front of the TV set every Wednesday for five years to watch a program with Morgan Freeman- Through the Warm Hole!

Nine years later, when Mr. Musk decided to create his new company Interlink, she began to scrutinized deeply about the secrets and mysteries of neuroscience. After the death of her father in 2013; she ponders on the curiosity to learn and know more about the nature of age and our ability to think and remember things clearly.

She wants to know how the brain has an innate capacity to create neurotransmitters not matter the age. After writing one book about How to use the power of the mind to control our thoughts; a second; about Poetry; and a third about Dark Energy; she decided to embark into a new discovery of the conscious mind, and the new innovative ideas about Artificial Intelligence interlace with the brain.

Frances believes that we should not limit ourselves to learn about one subject or use one idea to implement into knowledge but rather, she beliefs that we should expand our curiosity to discover all there is to know about us, the universe, the powerful influences of he unknown, and the unlimited power of the mind.

ARTIFICIAL INTELLIGENCE, THE NEW COGNITIVE MACHINE

HOW MUCH CAN we improve the thinking of an artificial brain? If it is true that we design our own reality, why wouldn't it be true that we can predict the future of artificial intelligence? First, let's discuss what AI is. Artificial Intelligence is the programming of information into a machine to artificially compute and process information visually or with a brain, allowing speech recognition, decision-making, mathematical calculation, and translation between languages. When we hear all these definitions, AI sounds far-fetched. The reality is that AI has been part of our system for years now. The automobile industry uses AI to build cars. The banks and other financial institutions have been using AI for years… Neuroscience has incorporated AI into a brain to translate language for the language-impaired like Stephen Hawking, the theoretical physicist from London.

The question is not whether we can do it, but rather, who is going to benefit from the strategic advantages of using a complete AI to achieve world domination. Will it be controlled by private entities, the government, or global agencies? If we move slowly, it might take years to create one, but if we allow competition to take place, the process of building this system may accelerate. Could one project have advantages over the others? And how do we control the speed or acceleration of projects? Who will be the main decision-maker?

Currently, the development of AI is publicized and open to all; nothing is hidden from the many projects. Nevertheless, there still competition between projects. Like with any other promising futuristic development, there is fierce competition between companies, such as Facebook and Neuralink. Who will have the upper hand? Superintelligent machines promise immense financial gain for the best project. If one project takes off secretly, who is going to have control or the final say? In open AI, any agency or private sector with the most financial advantage to control a design will have the power. However, if one project wins over the other, who is going to determine its value or reap the profit? Many questions remain unanswered, yet the idea is to make AI an open project for the most creative and inventive system.

Furthermore, we all know this is not the case today, and many are wondering what threats AI could pose in the future. There are those who highly oppose the secrecy of any project, thus paying a high amount to purchase all information about AI. Financial tycoons already wish to control the ongoing development of AI. Although it is supposed to be an open forum for all participants, this is not the case.

If we go back to ancient times, even during the time of Jesus, we realize that bureaucracy has always had the upper hand, controlling everything. Evolution has taken place; however, the old paradigm hasn't changed. Moreover, the more we try to control any project that has a promising financial future like AI, the greater the possibility of having a global market violation or unlawful contract for AI business. These projects can be sold at the highest price in the market to buyers interested in new AI innovation for a very large profit.

Moreover, this will demand a new government protocol to prevent or regulate growth within AI technology to prevent it from becoming a global threat.

Now, instead of having bureaucratic agencies control AI projects, the government will regulate how any progress with AI technology advances. Although the competition should not be hidden from the public, controlling when or how technology advances will certainly increase the disadvantages of global competition or outside development of AI. Stopping competition would affect who takes the upper hand at improving with technology. Hence, progress with AI is inevitable; anyone with a new project in mind should make it public first. Hiding any information could create a disadvantage for any project controlled by a monopoly not under regulation. Therefore, all competition or new projects would only compete for the best with or without the first public information. As a result, no information would be subject to control or regulation. More importantly, this system of publicly sharing information would be beneficial to the innovators because it would protect them from any copying or comparative disadvantages. It would be a mistake for any followers to copy or get ahead of the game unless they have made their ideas for innovation public or registered them first. This gives priority to the frontrunner. There would be no trade secrets. However, there is a great possibility for explosive growth with any AI project. Also, it is possible for the leading project to have a strategic advantage even if it takes a long time to develop.

Once AI takes advantage of labor, technical evolution, or energy to sufficiently administer all the production, the next phase will be machine intelligence projects operated by humans manually or with computers. The world will then enter a new era of job markets for the next technology. Undoubtedly, this new generation will surpass any today. Unlike with Silicon

Valley, new technology will not necessarily be required for the use of high-rise buildings or eccentric designs to produce a great environment. Instead, technology will generate a vast source of data programmers whose sole purpose will be to monitor all central systems to ensure AI is processing data accurately. Many programmers would only be required to check the system periodically to ensure proper communication with AI, in which case, an office would not be used. One data system here in the US may have consistent communication with another in a different part of the world, such as Australia, Vienna, London, Turkey, Africa, or the Middle East.

A central global AI might be a possibility, with diverse utilization of data around the world. However, there are pros and cons to the idea of global interaction. For example, there could be an attack or information that might be shared with uninvolved parties. Or we could have the best global communication ever established between all continents. The choice is up to us to design a well-balanced communication.

Nevertheless, we should consider that open-door communication is also vulnerable to some form of breach or opportunity to break the rules by those with malicious intent. Can we possibly imagine or understand why AI could be the most profitable technology of the future? The future of this technology is not only profitable, but also very beneficial to private as well as government institutions. There are ample opportunities for the first innovation of an AI that can transform society and us with one single idea. Similarly, without the best protection, we would be interacting with the enemy. Communication is essential for networking or global understanding. However, in dealing with adversarial countries, we might be stepping into the line of fire.

Nothing could be more critical than establishing a good rapport with countries whose moral, political, and religious boundaries can be at odds with our intentions to form a solid or healthy alliance. If we find ourselves in this predicament with nations opposed to our new technological ideas, we could be setting ourselves up for war rather than good communication. On the contrary, if we can create global communication, thus creating new, innovative ideas to help improve all humanity, we can create the best form of global network: an exchange of ideas to improve the entire world.

Technology is not only a source of network intercommunication but also a means by which large and complex situations can be solved in a short period. The purpose of AI is to use the most efficient intelligent to improve the quality of our lives. While some might see it as a complex measure for the future, it is up to us to determine what the motives for using AI are. Nothing could be more catastrophic than using intelligence for control or power to influence the masses. It is possible to use AI for a political agenda. Such intent would tarnish all possible outcomes for AI progress. In which case, every invention or use of artificial intelligence would have to be scrutinized beforehand.

There will always be an attempt by those with curiosity to try and interfere with new technology. At a global level, it will be difficult to control the network unless we make each nation or country responsible for their own actions against the system, and that will apply to all, no exceptions. The impact that artificial intelligence can have at a global level will be almost miraculous. If this is not done in the interest of any political purpose, but in the interest of the people itself, the increase in education, health, and welfare for millions around the world would be unprecedented.

Nevertheless, there is a great need for environmental improvement globally, and AI could help us change that situation. Finding solutions, creating new ideas for changes, or transforming how people are allocated to improve living standards with a healthier environment are several ways we can use AI to solve our problems. AI should be the future of intelligent solutions and problem-solving at a higher technical and intellectual level of thinking.

In viewing the future of AI, not all enhanced intelligence will be from a machine. We, too, are going to improve the way we think with the interaction of network communication when our brains interface with machines. This will require us to be more technologically in tune and think at a technological level rather than human; that is, if we want our brain to communicate with a machine. Our brain will have to function at the machine level until better communication has been established between the two. Once a good rapport between the machine and the human brain is created, the results would be effective communication, both close and at a distance.

We should see AI as the new smarter-than-smart technology with ample potential to change the world we live in or end it. Although considered to be smarter than humans, AI would also react like a machine, not a true human being. Herein is where the true potential for an unexpected outcome is highly probable.

Will we blame the system of the creator of AI? Are we, humans, becoming less compassionate because our emotions are now linked to a machine that experiences nothing emotionally?

Are we becoming more and more desensitized? If we continue this attachment to our gadgets or electronic toys; our human touch and compassion will completely disappeared. The only thing that will keep us from losing our human touch will be to connect to our mental, and our emotional conscious awareness. Consciousness could be the only cognitive transformation keeping us from becoming machine-like robots. Programming a machine is a tedious process in the beginning; however, once the progress has begun to function, it takes off, its knowledge growing exponentially. Only a signal from within an interference can significantly affect the system to vary and change its behavior. In this case, the coding will transform the analog signal into digital, changing its volume, thus resulting in a big internal immune virus in the program.

Consequently, degeneration of the signal will occur, resulting in degradation and manipulation of all the electronic components. In this case, communication could impair the system. This could be a great example of how machine intelligence or communication is the result of a programmer's input into the system to consistently obtain a good output.

It is possible for artificial intelligence to emulate us and learn about our behaviors, reactions, and impulsive responses and then inherit what it has learned from us and use it when appropriate, such as when it feels the need to act upon threat or in response to unacceptable behavior from others. We create every reaction a machine can learn from us by inputting information. Thus, we forget that a program is like any other data or information in space and with energy: it, too, can improve with time and practice. Not only can AI learn what we teach it, but it can also learn to reprogram what it has learned into a better method to use on its own. Learning requires energy, and a machine has a magnetic field of energy that improves with time and processing ability. Information is part of the field of energy or frequency that navigates throughout the universe; thus, data can be available to us. Just as we perceive knowledge and ideas and create from the frequency of energy in our neurotransmitters, so does a machine with input intelligence gather data and arrange it in a way that seems impossible to us. This energy from the neurotransmitters is made of photons with light reflecting back light into other neurons; thus increasing more energy in the brain that can be transmitted to a machine-like AI. Nothing is left empty between a transmission of thoughts or any form of communication between the two. Because energy is invisible, it can move through the field undetected. In this entanglement of the two, the energy of one affects the other. As Albert Einstein said; the field is the sole govern agency of matter. This means that at a physical level the brain and machine are exchanging energy we cannot see but is being transmitted. If atoms are a vortex of rotating energy and energy is invisible, at a quantum level the brain and the machine are entangled into a mutual vortex of energy they both share together as they are interlaced together.

Machine intelligence is but a copy of our own intelligence transfer into a system. To obtain superintelligence, brain and machine must intertwine. Anything we create with intelligence will imitate us, thus create more intelligence. Intelligence is not necessarily a function of the brain, but rather, it is a frequency in the field of energy that emulates and multiplies over time. It is nearly impossible to overestimate machine intelligence expansion. We cannot account for all the possibilities nor outcome of merging with machine intelligence in the future. What could happen during the process of intertwining intelligence with machines is still unpredictable. What we know today is that machine intelligence is not restricted to learning above the level of input; it is capable of intelligence expansion. Because all the body needs is a signal to activate its energy, and machines are powered by energy, the activation of the two could be a magnifying experience. Perhaps what will surprise us is how the neurotransmitters will respond to the constant activity created by the power between the two. The only concern we would have, is, if the program slows down the brain activities at the end of the day or at night.

What will surprise us one day will be the fact that AI intelligence will surpass our own, thus creating a system of data to implement new rules and regulations for us to follow. In this case, we will have to abide by its commands. We already see this happening with computer-programmed AIs. Such a system is monitored by the company that programmed it and can shut it down or disrupt it as they wish. It will be nearly impossible to know who has done this.

Consequently, the company or programmer may blame the failure on the program, not on the actual cause. This is only one of many possible cases of manipulation of programmable computerized systems. Nevertheless, with new artificial AI computers, the control of any manipulation can be easily detectable.

Nothing will bypass a program with AI that is designed to follow leads and determine where viruses originated from and how they have affected any program. Even if the virus has come from a remote location, it will know precisely where it is and how far away. Computers of the future are already functioning at high capacity to do this. However, the new AI super-computers will have greater safety than we have today. Due to the competition in new technological advancement, we are going to find ways to prevent all the gaps we have today with computers hacking and information violation. Understanding the nature of superintelligence in computers is not within our logic; not yet. We are referring to faster than light speed information gathering and accumulation, both. This translates into; nothing will leave any sector without a trace.

Some of the programs are being used behind the scenes. Not everything we see and understand is as it is. AI can be easily used for coercion, and it has been used for that before. The danger of intelligence lies in the use of it and the purpose it is being used for. It is not the machine we fear; no, it is us, and intelligent machines can do what we are not willing to do with integrity. Remember, any failure can always be blamed on a machine. This will free man from accepting the responsibility that he is one hundred percent responsible for the outcome of any programmable outcome from AI. Augmenting machine intelligence to combine with the brain can have implications such as; manipulating emotional state of mind control by a program. There could also be programs design to terminate any of the two subject, man or machine, if not longer needed. What if something goes wrong and the experiment between the two lead to highly private information lick out to the human? Will the humanoids be brain watch; or would they simply be redirected to another psychological transformation of the mind to erase their thoughts from the mind. What if the humanoid understands the system and the program both well enough to manipulate it, and change programs as desire? Then what? Some intelligence system consists of intelligence parts that are themselves capable of reinforcement learning. I wouldn't be surprise, if a humanoid, because of his entangle capacity with our brain could connect to a high human resonance working together in harmony with other intelligent human, and non-human species to activate higher frequencies. Theoretically, with any synthetic intelligence we create, everything in the field of conscious intelligence is possible. What would happen if we made a humanoid the best version of ourselves; one that was better than us. A humanoid with more caring and kinder capacity than ours. Then, their future will be better than ours.

WHAT IS POSSIBLE WITH AI?

LIKE ANY OTHER program, there is a potential for failure or that it is malignant. This also involves any possible catastrophe. This program could fail in various ways, one of them being a failure to respond when commanded or simply not having the opportunity to try a new program ever again. If we are going to create superintelligence in a machine, it must be able to help us solve any failure we encounter along the way. But in creating and testing artificial intelligence based on science, statistics, and data, we should consider that smarter machines are more dangerous than not. The logic is that if machines can solve their own problems, why do they need us? Scientists and researchers have been working to create greater-than-human intelligence for centuries. We have gone too far now to stop the progress of AI. And these promising new technologies in artificial intelligence are exciting and challenging for those who are working on creating superintelligence. Many large industries in technology are working on robotics today with positive results. Other vested interests in technology come from large investors and the military as well as national, economic, and security interests. Some scientists have built their careers on robotics applications, with other advance systems planned.

Artificial intelligence has a promising future in that

it can be more creative with intelligence and become more reliable with time. The more information or data we create to teach the AI, the greater the chance of creating a superintelligence. However, the importance of having success with superintelligence is to have those involved use judgment and ethics to significantly have the best result with AI and its function in the environment. Otherwise, we could see what many are afraid of, the machine becoming a superpower intelligence that can manage, manipulate, and take control of us. Imagine if AI had the upper hand at understanding how our security or defense systems work. If a machine can figure out what to do in an emergency, it can also attack us when it feels threatened by us. However, if an interface is active at any time, the machine can also tell the person what to do and ensure that he or she follows its command. This example is a hypothesis that could become a reality at any time. Machine intelligence can take over us or command us to follow their instructions to take control of any given situation.

Machine domination is something we should worry about in the future with artificial intelligence. Once machines' superintelligence has surpassed ours, they could take control of all situations, and we could become their slaves. We could depend on them for safety, security, survival, and so much more. If this were to happen, we would have no way of stopping them from taking over with a sophisticated global attack, one whose tactics would depend not necessarily on weapons but on denying us what we need to survive and making us suffer. This could decimate the population, and it could even be the end of us.

Could an AI do this? Consider the possibility of superintelligent machines ruling the world with the power to do as they wished, all because we taught them how and when to act when we depend on them for everything. The danger lies not on the machine, but rather, on our ability to depend on them to do everything for us as we teach them some of our deepest secrets about safety, security, and war. The fact is, machines are not necessarily thinking devices; they are functional devices that learn how to do things according to simple but tedious instructions they learn. But once they have a handle on how things work, they become faster and more accurate than we are. Intelligence is not something a machine learns but something it practices and gets better at with time and constant repetition.

One of the fundamental elements of quantum physics is the ability to transfer and decipher information back and forth. Once a machine has learned how to transfer, copy, change, and exchange information, it can do anything with it. The possibilities are endless because everything can be manipulated; it's an entire system to do with as it wishes. The problem is not how much information is available; no, it is about how this information can be used to corrupt any program while we sleep, believing that all is well because AI has control of everything.

Although we see AI as part of our future progress, there is no doubt that with any new program, there are pros and cons. Until we understand the nature, purpose, and potential of AI, we cannot conclude what that outcome of developing it may be. It takes time and experimentation to learn to put all the pieces together and fully understand the complex nature of artificial intelligence. We must realize, though, that at any time, AI can act according to our commands or against them. If this were to happen, how do we protect ourselves against a superintelligence we have created but don't know how to control. Because of the complex nature of machine intelligence, we cannot be certain what the future holds for us with AI. If a machine becomes super intelligent, it will exceed our intelligence capacity. It is obvious that man have yet to exceede his unlimited potential for intelligence, if he is still seeking within an outside object to test his own intellect. Tesla said, "The continuous development in this direction must ultimately make war a contest between machines without man in it." As men continuous to seek for intelligence outside himself, he could be defeated by that same intelligence. Hopefully, we can think of all these intellectual gathering of ideas about the future as simple speculations.

Most of Tesla predictions and theories about the future are today a reality. I like to think about the future of us to include not only Tesla, but myself in it as well.

MACHINE MANIPULATION

WITH ANY PROJECT to make humans smarter, there is treacherous potential for danger. There could be an intelligence backfire, and the AI could become smarter than us. Or we could face the potential danger of having to make a strategic decision to act prudently or face the consequences. If the program is self-reliant, the outcome could be threatening. And if we decide to terminate the program and create a new architecture, the programmer could face the problem of having to build a new utility function to make it safe and more hands-on to control.

But what if this new AI has already learned the new ways to function under the commands but still can read all the data from its quantum intelligence to put together all the pieces of the puzzles and make common-sense conclusions about the previous AI? Then there exists the problem of machine intelligence being underestimated by the programmer. In the world of information, one supersedes the other, but traces are always there; information remains readily available if we seek for it. Even though we may have to terminate one AI to build another, the lesson is that we should realistically evaluate all possible outcomes with superintelligent machines. If we don't, we may discover how superintelligence can obtain power over us. It would be shameful to see that machines have finally outsmarted man and taken over.

Nevertheless, we seek an artificial intelligence that can invest for us, have personality, make us and our prospects happy, become intelligent enough to be reliable, find us pleasure in all areas of life, and be there when we need it. But this dependability on AI is what would force it to take complete control over our lives without reservation. If AI has a great advantage over us, we will have no ability to stop it, and our attempts to do so will fail. On the other hand, considering the level of intelligence of AI, it may trick us into believing that it will obey when asked to do so, only to turn and do as it pleases. AI will have learned to manipulate us and take control of any situation, not just us. All the intelligence resources will come from the amount of data input into the system or program. The results of its behavior will not derive from our actions, but rather, they will come from all the information collected as it was learning from us. It will put its own logic together to compile information and reason as it prepares its next plan.

We could get to a point where we no longer have control of AI and it is concealing its intent to create new ideas and act on its own. AI intelligence will develop to a point where we can no longer control nor predict its actions. At this point, we will not be able to control or destroy the AI because it will know how to duplicate itself with more intelligence than before. We would have then lost complete control of it and would be unable to fight back.

With superintelligence comes control and manipulation, the same as with us humans. What we are creating with artificial intelligence is a replica of our intellect with more powerful influence and thinking than we have. It's a level 10 on a scale of 1 to 10. The danger is that we create something to emulate us and that something then gains control over us.

It is feasible that intelligence is pure energy in perfect harmonious accord with the field of everything and, thus, it takes the largest field to amplify intellect to the highest possible level available in the universal field of everything. Consider how atoms or electrons navigate throughout space and then multiply themselves as they reconnect with one another. This constant motion is energy flow at the highest level in the universe. Once this energy multiplies, it continuous to amplify with other energy emancipating an entire sphere.

Like the world wide web, the transmission is transfer to multiple networks available and to other energy sources of energy that can be manipulated from any angle or location. Because we need a system of a machine to help us transmit the energy, and transmission; there is a gap or wide-open door for manipulation to take place. Consequently, when we have a system manipulated, it isn't a machine that we attack, rather, is an entire program or system at large. As a result of a system or program manipulation; it is the humans who suffer the consequences.

INTELLIGENCE EVADING

SECURITY MUST BE our priority if we want to stop the corruption of hackers trying to invade the privacy of any individuals. The more we think about protecting our lives and our devices, the better things are for us. For the security and safety of all individuals, artificial intelligence of the future would be able to read a person's mind through a computerized system and understand whether the individual is being honest or not. AI is not only capable of knowing what is going on behind the scenes, but it can understand through words and commands the actual thoughts of a person behind the computer. The intelligence will pick up only a few words to detect the truth or falsity of statements. We write as we think, and superintelligence puts words together that make complete sense when organized properly. How AI can read each individual person's way of writing has to do with the training that goes on while learning how the brain sends messages to the mind to then put them into words. AI will know exactly how a person feels when they write the note or letter, what is going on in their mind, and even how they are programming their thoughts according to their feelings.

This system reads everything, both in front of it, and hidden from it. It may sound like mental telepathy, but it is more the act of quantum communication. Once the intelligence has learned how to tap into the mind or brain to read its frequency or energy flow, there is no stopping it. Because the machine is also interfacing with a brain, it learns all the functions of the mind and puts together a perfect array of interpretation on its own to make sense, learn, study, and create a perfect conclusion as to what is happening now.

However, there is always the danger that those interfacing with the system will commit a violation of privacy, extracting personal data from others. If interfacing connects the brain with a machine, other systems can and will pick up either frequency or some form of information. There is nothing in space that cannot be extracted from the field of information. This is because everything that has been created and exists leaves a trace or a blueprint behind. Even at a mental level, this information is traceable.

The question remains as to how we protect our privacy once it has been installed into a system of the mind and machine to communicate mutually. Codes won't help. This is when we would have to be very creative to ensure that our minds are not hacked. Yes, hacking can retrieve information from our minds. In dealing with privacy, it seems as if we are constantly chasing a ghost. We must think far ahead of or like hackers to prevent them from invading our lives. However, there is only one way to make our privacy unavailable to the public network and prevent hackers from getting our data: taking it out of the system. The less we expose our privacy, the better chance we have at keeping it safe. Unfortunately, that is not the case today with our personal information. Who say we must pay a company to keep our records; such as credit records and personal medical records? Why can we keep them and only release them to people we authorized them to. Everything in our system today is pure manipulation. We must be intelligent enough to fight for our rights to maintain our own privacy. My personal information should not depend on the holding of information by private institutions using them for financial gains. The future of technology most provide a flexible program for individuals to have control of their lives, not the system. Freedom and liberty are not promulgated by holding others accountable for our freedom, but us, the individuals. AI will have to solve this inefficient way of handling our lives with a super- intelligence evading us from control. We are under the assumption that the government takes control of our safety and security. But do they?

However, government regulations barely protect us from having any privacy of our own. Every bit of information is made public with the system that is supposed to protect us. There must be a way to keep our privacy available to others. If we want to counteract invaders, we must think like them. We should become astute enough to fight back. This applies to all aspects of our lives. Superintelligence will teach us how to be honest, authentic, and truthful to ourselves and the rest of the world because it will define us in more ways than one by our behaviors once it has learned to tap into our minds. We should not fear machines or programs, rather, we should fear ourselves as we approach closer to discovering the magic of quantum bits and machine computations.

SUPERINTELLIGENCE AND COMPUTERS

SUPERINTELLIGENCE MEANS THAT a system or something can operate at a higher level of intellect than humans. This superintelligence must be two or three steps above our normal thinking habits. It should solve our daily inconveniences with great authentic precision to help us live a better life. Then it should be twice as smart as the average high-IQ individual.

AI should also be able to detect any security infraction coming from another system, whether it is local or long distance. The system must oversee and override whatever is coming into a program by nature of its frequency. Consider this: when a person is trying to hack any computer or data, they try consistently until they get in. Now, the rate of speed gives AI the clue as to what is going on, and it can automatically detect that interference is on the move. It will immediately respond with an alert to a data center with a signal that indicates there is an intruder or hacker trying to invade the system. Then AI will make a complete background check of who is the user of the computer, what is the password, and if the password is being hacked from a distance. Once it has found the code, it will send the data to the main programming system, and it will immediately know the hacker's location and actions, and nothing will be secret anymore. With this new coded programming, AI can act immediately, and there would always be a trace back to who has done the hacking.

If we want to have a superintelligent design to communicate effectively, we must program it and evaluate all data with precise accuracy. Once this process is active, there will be no more data hacking from private or individual sectors. Security will be ensured, and we will have privacy. This will be done in cooperation with the new program, which is now in use and up in space, keeping an eye on us. AI will evaluate all content together; then it will communicate back to us. This data would be like a calculation but without errors. This will make its deduction and evaluation incredibly precise and accurate.

Did you know that today there are new satellites in space keeping track of our entire lives? Our lives are constantly being monitored, not only through social media, but also via these satellites. We do not possess as much privacy as we think we do. However, that is about to change with AI. A new and advanced intelligence that interacts with our brain is about to transform the way we think and conduct our lives today.

Just as we think that life is getting better, so are we going to be gradually seduced into a new way of behaving and conducting our lives. The idea that a machine will arbitrarily change the way we live and do business sounds like sci-fi. Our fascination with the future and artificial intelligence will take us further than we thought, and there is a price to pay for it. Because NLP/neurolinguistic programing is part of the programming in AI, the way we communicate with a machine will also change the way we use our brain for thinking and communicating. If we should incorporate a machine into our daily routine, chances are we will also have to change how we interact and do business with each other.

The future does not only belong to us, but it also incorporates how we do business, how we implement our ideas into new possible business planning and programming for the future. Once we have taught a machine the rules and regulations of a business or program, we will have to follow as it has been written. It is not possible to set up a data system for a machine to run and for us not to follow it as designed. This is where AI will either transform our lives or make it difficult for us. The idea that we can design a system to perform for us without us having to follow the same rules doesn't sound reasonable. As we make our future more interactive with machines, we are bound to respond and follow as we have programmed the system to be. As we can see today, must of us are already in tune with a device in some form.

Let's take it to the extreme as we objectively see the future from the bright side. Life will be more enjoyable, more relaxing, but also a bit stressful. But how about those who are not in communication with what is happening in the future? Where do they feed in? Are we going to segregate society into those who know and those who don't? Will the future with machines be interactive with some, but not all?

We can predict that the only way to topple a totalitarian environment with change is to educate and program the minds of the many. We must accept changes in the future and ensure that all humanity is part of the plan. In such a case, the cost for educating the masses could be solved by using those who are in a program to educate the rest. The unfortunate truth is that a system like this is not available today. The cost of living is rising, education costs more, nothing is given for free, and we are not all equal when it comes to education. How can AI help us solve this dilemma? If we look at the future of space travel, we know with certainty that only those with means to engage in such a dream will benefit from it. All else will be left behind.

When it comes to our present reality, nothing is simple. Reality gets complicated with time, finances, or the level of knowledge of what is happening or who is involved behind the scenes of our daily world. Our world is being manipulated at every level, so it will take an ingenious machine to desensitize those who are deeply involved into having control of it all. The question remains as to who will design a system that will change or transform what is going on today and reprogram the minds of those who have the power to control the masses. Even at the level of machine intelligence, those in control of the system have the say. This world would be a better place if we had a system that monitored all outcomes according to the regulations we have set in place. Perhaps artificial intelligence will one day help us solve the enigma of power and control by the system rather than bureaucratic voices determining our lives.

Today, the truth is stranger than fiction. We don't know to what extent our lives or information are being scrutinized or who to trust anymore. It is not a matter of doubt, but rather, a matter of deception by those we entrust our lives to and those who are supposed to protect us. No wonder we need a system like AI to oversee our lives and create some order. I believe that this is what those in power fear the most: they fear a program will expose them and all their hidden agendas will finally be known by the public. Perhaps order is what those who create discord are afraid of. What would happen if AI discovered all the hidden agendas going on by those in power? How would they face the world, what would they say, how would they act, and what would happen to them? Can you visualize this for a moment?

I would love to see what takes place once we all know the truth. Will it finally change the way we see the world, what we believe in, or will this finally give us faith in something we unpredictably plan without thinking about its outcome? In programming any system, we must think of the consequences behind it. We cannot possibly imagine a world without a counteractive response. Everything has a balance that eventually brings harmony into place.

A good hypothesis would be $E=MC^2$. Energy equals mass, and mass equals energy. They counteract each other to bring a perfect balance. Without it, there is no harmony in life. Such is the nature of everything. When there is an unbalance, there is a negative vibration of energy flowing. Likewise, in a system without control. Life is a complete balance; without it, there is no growth, no harmony, no peace, and no advancement. Such unbalance causes disruption in all areas of our lives. Our present world is not in good or steady balance. The idea that a machine can change our lives for the best is not a mistake. The need to restore and balance to the scale of our lives needs adjustment, or we will reach the point of breakdown.

How do we balance the progress of AI, which is moving faster than we had anticipated it to? Although we might consider AI to be in a stage of infancy, others see it as an open door for the future expansion of society. Thus, to what extent do we allow the expansion of AI without having a negative outcome? In addition to setting rules and regulations to control its advancement, AI innovators should create a panel specifically for debating whether new projects should be pursued. However, if this panel is formed in one region, there would be other regions where regulations do not control the advancement of AI. This will make it possible for new AI technology to be implemented without further approval. Consequently, no matter how much control is use for AI technology, there will always be someone seeking to do otherwise.

For example, cryptography is a secure system that embodies software transaction codes with a private identity structure locking personal information from the public. Blockchain is one of them. This system is done with quantum code. We tend to forget that quantum has many probabilities; therefore, it has ample active streams of transparent layers of options that can be broken. What this requires is a tracer that can read codes. Although blockchain changes with consistency, tracers can find the changes and form a pattern of numerals or codes. Tracers can read the codes in the air and trace it back to a system to read. That which is quantum can be traced with similar data, for it has multiple structures of possibilities. The only way we cannot trace back any codes is by using symbols. If the data is changed frequently, it can still be traced back, though. The only way to completely secure any code is to encrypt it into unusual symbols that only two recipients know and nobody else.

ARTIFICIAL INTELLIGENCE AND CODING PROTECTION

THERE IS ALWAYS a way to let the cat out of the box, for we don't know if he is dead or alive... If creative minds exist, invent, and think, there will always be a way for others to try and counteract or imitate what has been created. The problem is not with the system, but with the ones who create and invent the system: us! For blockchain to be completely secure, we would have to invent several ways in which to use codes, secure them, and consistently put them in a secure place where no one can access them.

Changing the pattern of codes is more important than shifting them into different codes. Codes can be traced, but encrypted symbols codes are difficult to decipher. Can you imagine using a flower category or description as a code? Who in the world would know the meaning but the one who created it? We could use a pet breed as a code in a variety of incorporated breeds. It would be very difficult to trace it back. Anyone who did would have to be a genius. If there is a slight chance for the codes to be deciphered, one code will immediately be disrupted because it has the quantum potential hidden in it to split the moment it is detected. The code will automatically separate and disappear. It would be like having a code with quantum superposition.

Quantum possibilities exist in the world of the tiny with magnified energy; therefore, information is everywhere, and it can be extracted if necessary. The fact that particles behave like waves and waves like particles gives us an insight as to what is possible when coding information or storing it in space. The outcome has multiple possibilities. Just like Schrodinger predicted, they have different probabilities, which connects them to the apparently absolute or deterministic nature of the large-scale universe that has given rise to many interpretations of quantum physics.

To cover up any traces, one will have to create an ingenious way to hide data. Using information as codes can cover up any traces. The essential factor is to not use numbers or letters; rather, we can create individual encrypted symbols into different designs. Although the code of the universe is deciphered with numbers, there are codes, such as the cells in our bodies, that are encoded by symbols. In this case, coding will be considered a secret or private source of nontraceable data.

The only problem we have is recreating a new code to replace the old. It will require us to act quickly and efficiently with new ideas to formulate a new secure code, one that might take a long time to find it. Codes are only traceable when we use a new code that is like the old. Therefore, making tiny increments of changes helps ensure security.

Blockchain is only the beginning of creating a safe way to keep our records in order. As we advance with AI, there will be ample possible ways to protect ourselves from the constant hacking of information. However, unless we use intelligence above AI, we cannot compete with those counteracting our intelligence. In the past, wars have been won because one astute individual used a different tactic to outsmart his enemies. In coding, one should think similarly. Hackers or intruders are the enemies of safety, and they must be treated like enemies of our safety and security. Unless we want to be dominated by machines in the future, we must think ahead of superintelligence. Creating any system, robot, or program that can exceed human intellect requires that a person or machine be smarter than us.

In the quantum world, the universe has given us many interpretations of the big and small. Quantum values use numbers to quantify values. If we use symbols with the same pattern, we are more likely to find a solution to cover up information from hackers or any possible deciphering of codes. In the quantum world of electrons, there are multiple possibilities. Numbers are not or should not be the only solutions to use codes to protect against hackers or decryption. There must be a pattern we can create to make it impossible to be traced by anyone who isn't part of the system created. If symbols, codes, and patterns cannot help us solve the problem of privacy, we may have to come to a point where the palm of our hands will be the only form of identification for transactions or even purchases of any kind. We are creative human beings, and there is always a way to solve any problem. Our capacity to reason helps us see beyond the present and, thus, consistently improve our lives and our society.

When it comes to machines, technology, and science, we have more knowledge today than we ever have before. Our interaction with machines has allowed us to understand the basis for creating and interacting with them. There will come a time when we are going to improve our intelligence so that it equals a machine and create ideas that will impact the world with a single touch. Machine, the mind, the heart, neurons, cells, and our DNA will impact the world we live in today with the understanding of invisible forces of energy from a quantum world of electrons and photons. Could our DNA have anything to do with the world of quantum energy coming from photons? After all, "Let there be light" is a form of interpreting this idea. If light began the universe, in light, all sources of creation are found.

At a quantum level, the world seems small and invisible, and the mystic and miraculous nature of the quantum world seems impossible to understand, but it is not. We can see that even during the time of Jesus, the act of invisible power, or what we refer to as the Holy Ghost, had magic in it. How is it that a man can levitate in midair and lift himself up into heaven? The invisible world of photons has all the power and secrets of this world in a tiny bit of particles with magnificent energetic power. No wonder that experiments cannot come to a single conclusion as to what these tiny, powerful electrons are or how they turn from particles to waves and spread or cancel each other out. Our lives are intertwined with the universe, and every invisible aspect of it, we refer to as miracles or magical. Nevertheless, when we connect with this magical part of the quantum world, miracles can happen.

This quantum world we refer to as invisible is the most powerful source of magical energy available in this universe. Therefore, at a quantum level, everything seems like a miracle of mystical nature, not easy to understand. When it comes to the world of the unseen, the magic is not in the physical action or visible appearance of the act, but rather, it is inside what we cannot see but sense with our mind and spirit. The word "magical" means power of the unknown or the nature of an invisible power we cannot see with our eyes but can sense with our body, our mind, and our spirit as it transitions around us or through us. Imagine a powerful laser traveling past us at faster-than-light speed. As it travels, we cannot see it, but we can feel its presence, if only for a split second.

The word "quantum" refers to magic, but not in the sense of evil nature but rather in the aspect of invisibility or something with power. Energy, power, and light ignite everything, from our heart's neurites to the cells in our body to the magnet in our bodies or the nerves that connect every single cell with the brain. We, too, are quantum creatures of this universe. Everything that was created to give us a comfortable life here on earth is also part of who we are as human beings. All aspects of creation are in us and with us in every living, breathing cell of our bodies. We can understand this in our emotions. When an earthquake happens, we, too, sense the emotional vibration of the earth in our hearts. When rain falls from the skies, we, too, have a sense of soothing feelings, and as the sun shines over the entire planet Earth, we, too, sense its rays bathing over us with powerful energy. Every living aspect of this universe is in us and with us as it transitions throughout time in space. There is no separation between us and the universe; therefore, at some point, we are going to get the full understanding of our relationship with everything that is in this magnificent universe of ours.

The more we connect to the world of energy, magic, and power, the more we will discover, learn, and understand ourselves and the entire universe. Understand that magic does not refer to the aspect of evil. If that were not the case, then we could not refer to Jesus as the Holy Ghost, for in him, there were magical aspects as a human being. According to the Bible, Jesus is of the quantum world of magical, powerful energy in the universe; after all, he lifted himself up into the infinite, and he lived among humans, didn't he? The world of quantum is magical, it is power, and it is energy, but most of all, it is invisible. The mystic of the quantum world exists in the aspect of a non-physical world of magic of the unknown. It is a tiny world of powerful energy that extends into the infinite in a way that we cannot see nor experience with our eyes, but perhaps, maybe with our consciousness, we can.

WHY PROTECT OURSELVES FROM HACKERS?

AT SOME POINT, hacking should be considered a form of treason. Maybe AI will create new rules we must abide by and follow as we engage in brain-machine intelligence. Imagine a future in which AI will detect a person's DNA at a distance once it has been recorded as part of a personal file if this person happens to hack a system, break the law, or simply commit a crime. The only way to ensure safety and security is to make those responsible accountable for their actions. At some point, security from intruders should be part of our inalienable rights. Living in a peaceful world should be a privilege, not a treat. The best gift evolution has given us is the opportunity to advance and make ourselves part of that advancement with a more intelligent approach to life, safety, and conscious awareness of our environment. We are responsible for what happens with new technology and AI, how we program it, and what the outcome of that program is. In the end, we have the upper hand to make it a good system or make it disrupt the system we create. The outcome of using AI in our society or system depends how the system is used and what the benefits to society at large are. Without us, there is no AI.

Maybe AI will be the first technology to invent marriage proposals encrypted into a code and voice recording from both parties to ensure safety from fraud or malicious intent. Or all documents will simply exist in a personal file created for everyone, with DNA, blood type, and fingerprints, no dental records needed. AI will create a new world of billionaires because its invention will produce more safety and security for the future than we have now. Having a system in place that gives us peace of mind is very important. Because our privacy has been violated in the past, new and improved inventions will ensure that we take control of our lives and keep our records and lives in order. AI will provide that and more.

What if, instead of fighting an internal and external war with hackers, we programmed AI to learn how to track any hackers regardless of the distance. Whether it be a code, or a link AI follows, AI will find it. The fact that quantum mechanics and communication at a distance is not a fallacy anymore; gives AI the opportune time to help us solve the riddle of hackers and our privacy. We are about to enter a world of computerized programs that can help us detect anything that is been created by man. We are going to need the help of an AI program to solve our present problems with hackers. Hence interlace will help communication between man and machine more efficient; communication at a distance will create a new way for us to have information as needed. But first, we will need to create a quantum computation computer that relates data at the speed of light. It will take a very tiny field of the infinite small atoms to optimize with quantum bits microchip to solve our problems.

Meanwhile, quantum communication at a distance or spooky reaction as Einstein said, will help AI relate data to us that can help us track any information obtain from hackers.

The future of artificial intelligence is not only for use of high tech; it is preparing us for a world of changes we have yet to imagine. Is preparing us for the magical world of the big, and very tiny; quantum mechanics.

THE IMPACT OF AI ON OUR HISTORY AND SOCIETY

COULD IT BE that this time in history will be remembered as one of the most fascinating and unforgettable memories of the twenty-first century? AI will help our generation make an impact on technology and transformation in society that will be remembered by lasting future generations. Imagine all the things we can discover, invent, and create with AI. We are in the process of doing that and lots more.

However, AI is not the means to an end; the next highest impact on our society will come in the form of man's ability to do and create with his mind at a very intuitive level what he never thought to be possible. This will happen because he will tune in with a machine or system that will enhance his brain cells and activate the frontal lobe to its highest level for him to learn and process information better than ever before. This is possible because we can activate the neurons and rewire the brain to form new patterns of neurotransmitters to accelerate the learning process in the brain. Once we incorporate the brain with machine, something miraculous will happen; man will become an autonomous thinker. Man will know with his mind what a person in the opposite room is thinking and relate to him at a distance. We will be able to read or know each other's minds.

Dangerous or good? This simply implies that as our intellectual capacity develops with a machine, it will require us to change our lifestyles, how we do business, practice medicine, and even educate ourselves. The only question that remains is whether ethical values will be broken or will provide advantages over other AI discoveries. Although all information pertaining to programs in progress is shared with the public, there is always the possibility that somewhere someone is going to break the cycle of trust and violate the rules. Not all countries have rules or share information with the public as we do today in America, and that's where the difference in having the advantage is. Consequently, no one will be able to predict who or what will be the first to take credit for AI. However, the difference will not be the implementation but rather the benefits of its use over another AI.

To some, the idea of creating a superintelligent brain sounds diabolical. However, it is not intelligence we need to worry about, but rather, how such intelligence is used to manipulate or control others. The purpose of intelligence is to be used for discoveries, for advancement and improvements in technology, knowledge, and all other aspects of our lives. How will technology impact our world if we don't control how it is used or implemented? Anything that is out of control cannot possibly give us a good result. Eventually, the forces of nature, whose powerful energy always flow according to a perfect harmonious balance, will fall into place, and when they do, any incoherent disruption will fall to pieces.

It comes down to the perfect and impeccable equation by Albert Einstein, $E = MC^2$. In this equation, we find the perfect formula for harmonious balance. Mass = energy, and energy =mass. What this implies in simple terms is that life is a perfect balance in which all things must work in a harmonious way to achieve excellence; if not, the results are out of sync with your decisions, your purpose, and your meaning for existence.

The progressive world of ideas will bring about a new era of medical technology for reading and scanning the body to detect anything lacking or debilitating with our health. Thus, this technology will provide all the necessary balance for the body to return to its healthy, normal state. It will also determine the longevity of the person based on DNA, blood type, and even cognitive intelligence. With this technology, we will study the cells to understand their health, and we will learn how long they will remain in a healthy state or when, depending on their present state, they will begin to decay with age. However, as they begin to decay, new cells can be administered to keep them in a healthy balance. Such a process of healing can be done with stem cells or a machine that repairs cells by destroying the old and creating new ones instantly: a cell-regenerator machine.

Can we do the same with a heart transplant? Can one imagine the thought of waiting in the hospital for the system to create a new heart? If we are going to imagine the future, we might as well get creative with it. Also, we can imagine a future where eye transplants can be as easy as choosing what color we want and where the new eyes can be sure to provide 20/20 vision for everyone. Why 20/20? A new device will check your vision, provide you with its status, and determine the correction needed. It will then program this correction into a system that will give back all the details necessary to make the correction; It will also tell you how soon you will be able to see and how long this new vision will last. It will be permanent. Thus, quarterly corrections will be administered by the machine or system based on your visual details. No doctor will be necessary to perform this task because the new machine or robot will be equipped to do it all for you. This would be a complete visual imaging and profiling for the eyes with permanent correction.

Is it possible that we could do the same for the brain? Study the areas of the brain in any human being and try to correct what is missing or is not accurate; this is the neuroscience of brain enhancement. This will, of course, require a deep study of the brain, its cells, the present state of its neurotransmitters, how and why it is functioning the way it is, and what we can do to help interaction with the cells or neurons. How can we use any device to help neurotransmitters to accelerate, improve, and communicate with each other more efficiently?

The possible outcome from blindness will no be far from a reality with AI. We must find the way to make it happen. Today, there are many who opposed to this idea; except me. I dare to think of the possibility that one day the blind will see.

The stimulus color begins in the retina as wavelength gets higher processing from the visual cortex in the back of the brain to the cells. All the cells together respond to the visual cortex. The color perception processing occurs at the retina where five types of neurons correspond to pair colors to the visual cortex. They are V-1, V-2, V-3, V-4, V-5. These send more visual information to the brain. There is a perfect intercommunication between them. These nerves are like optic fibers, and they send signals back and forth to the optic nerve and the retina.

Now, if you could imagine A T.V Screen, A Camera, A Screen on the Computer, they all have a reflector that sends the images to the screen then become visually present, correct? If we were to insert a microchip inside the back of the brain in the visual cortex; then insert a pair of lenses in the eyes with a tiny camera inside of it. There is a great possibility that if the nerves cells in the eyes are healthy, we could obtain visual images from the visual cortex to the lenses in the iris of the eyes. As the nerves are stimulated with signals from the microchip to the retina from the optic nerves; optic radiation is sent to the cerebral hemisphere then to the occipital lobe. Then, finally, this radiation reaches the visual cortex. The visual cortex will then begin to process conscious stimuli to activate the modules and send signals to the iris and the lenses. The lenses will then reflect all the images to the eyes. As the patient receives signals to the lenses, vision will become activated. For a patient with vision for the first time, the images will be confusing, but with gradual training and practice, like a new baby, the patient will begin to learn every step of his clear vision as something new. It will feel like walking for the first time.

At first impression, his vision will feel like using peripheral vision at night, in the dark. Then eventually, adjustment to the camera will be necessary for the patient to learn to interpret images, forms, colors, and understand the meaning of a brand-new world. The act of seeing doesn't mean the patient will begin to act normal, he will have to adjust to his vision as a stranger to a new world of images, forms, and colors unknown to him.

Here is another idea, if the eyes cells and nerves are damaged or impaired completely from blindness; we could do an eyes transplant were the eyes are healthy and use them with the microchip and/ or tiny camera to try to make this happen. What are the adds that we can make this happen? We have invented hearing aids; we do plastic brain surgery and we insert all kinds of metals into the body; why can we transform eyes vision for the blind? I believe we can.

WHAT IS AI, WHY DO WE NEED IT?

MANY BELIEVE THAT when the new age of artificial intelligence begins, they will not be defeated. Perhaps this is what many fears about AI. To understand what AI is, one would have to think like an AI. It is but an above-normal intelligence or a machine whose intelligence exceeds that of a human brain.

Our irrational understanding of what AI is comes from the notion that it is simply a machine, or a system organized by our own programming. However, we have failed to recognize that machine learning doesn't necessarily come from programming, but rather from the consistency of the program to constantly reprogram itself as it runs over and over for a long period. Could we say that machines do develop intelligence because they run 24/7 or simply because they are learning by practice? There is no difference between machine learning and the way we learn. Although we may use logic to rationalize things, a machine does not need logic to follow; it simply puts into practice what has been programmed.

On the other hand, if the program isn't available, AI will follow precisely what is before it. Think of it as dummy practice. It simply follows; there is no algorithm, no special instructions or program to follow. Can we infer, then, that once we have created a program for AI to follow, we have created logic in the machine? In some form, we have. The logical understanding of how to program the system has been automatically transferred to the AI, which now follows instructions according to our guidance and logic.

Now, how will this play a role once we interface our brain with a machine? How efficient will the response be in communication? At some point in the interval of mutual communication between the two, there must be some logic transfer to AI for it to learn how to follow instructions from the brain. There should be a variable that helps us establish a great rapport between the two if we want communication to be successful. If AI were able to read our thoughts, what would be the complications or benefits? Would there be secure privacy of interaction with a machine to protect individual privacy, or would it be a matter of us being careful about what we think because it could be public information? No matter what, once the system is programmed, there must be some protection for those individuals who are going to take the risk of connecting to the device or machine. All personal data cannot simply be displayed to the public. It is my firm belief that the attempt to connect to a device will be for the mere purpose of personal privacy or gain.

Why are we, creatures of advanced intellectual capacity, seeking to interact with a machine? Is there a reason for this? Why connect the brain to a machine? Are we not intelligent enough to cope, think, understand, or even use logic? Why interact with a machine? What purpose is man seeking to personally be attached to any device unless it is for the act of seeking better health? Perhaps we need to look at the advantages of this interfacing. If we have enough reason to do this, then we can say that it is worth the effort. However, if it is for simple curiosity or even ego, then the dangers are greater than the benefits.

Whatever the case, at the rate we are advancing with technology, any inventions or discoveries are but challenges for the new generation. We have reached an era where thinking and creating are advancing at the speed of thought. Therefore, we need to improve the way we live and make a big impact so that the future is better than the present. For the future is now, and we are living in it as we speak. Perhaps what we are seeing now is but the beginning of a new dawn, glimpses of what is coming. AI has sparked curiosity, and its potential for benefits is unpredictable thus far.

Because the potential for financial rewards is so great with AI technology, the competition, hackers, and opportunists will break all the rules to get ahead of the game. I am almost certain that this is the case today. Stealing ideas and information has done massive damage since the invention of the internet.

Browsing for data is the new way companies, educational institutes, and private sectors find new ideas to implement into their ongoing projects for AI or new technology. The World Wide Web has made it possible for this to happen. On the one hand, it is the best source of data, but on the other, it tends to allow personal data to be leaked to the public. However, there is always a way to deceive the deceiver with information and trace all the data back to where it came from. Here is where AI can be our greatest tech spy of the future, a subagent, or an autonomous agent.

WHY ARTIFICIAL INTELLIGENCE?

THE INVENTION OF AI is not only for human advancement, but to bring with it a certainty and security that can never be violated. Information is one of the keys to ensure safety in a technological world. And it is out there everywhere. The focus on AI is to process information data for us and output as we need it. Therefore, AI can also be the best way to secure safety and information. Now, because AI seems to be the best way to protect the public from hackers, it might also appear to have control of our privacy. Hence, the intentions of a machine are unlike those of humans, so it is better to trust the machine than anyone who wants to steal for self-gain.

Let's put into perspective the truth about hackers. They may program the system to hack another, but it is the system that follows the code and irrevocably finds the way to track that code after endless trials. Without the machine or system, the hacker is nothing! He or she cannot track a code without superior intelligence. In this world of information, nothing is private. If we consider the value of quantum mechanics and particles that split but communicate at a distance, we should also consider the possibility that information is everywhere and can be tracked no matter how it is stored or preserved. Because particles travel at a distance, other particles of similar nature can also communicate at a distance, thus create interference in the code or communication network. We would have to be more creative to establish safety and security without hackers or interference of code breaking. Like every other system, there is a process to follow to get over the hurdle and find the core of any problem. Eventually, the problem is solved, and we should become more creative and imaginative to stay ahead of the game.

Every process has incremental steps to follow. Creativity and imagination are the way of the future. Your mind is a fortress, and you are the guiding force behind all your thoughts and creations. What hackers do is simply use a tool to help them navigate through a network and use data to their benefit, but what they don't understand about information is the nature of the quantum world. This is part of the nature of the universe: nothing is hidden, it is there, and it, too, can be found. Why artificial intelligence? The question is why not? We can use AI for protection against hackers, for our future, ourselves, global interaction between countries and political or warfare; the list is endless. AI should not be a threat to humanity nor a financial gain to benefit those with bested interest in our personal goals; AI should be what the secret service and national security are, protection for all. But are they? Is that why we need an AI. If we need an AI to protect us in any area, perhaps, we lack the safety and security we deserve to have. That's is why we need artificial or synthetic intelligence to protect us. I have to say that, we need AI not so much for national security but rather, we need it protect our personal identity, to protect us against domestic and international financial frauds, and finally, against the epidemic of the ill minded hackers stealing data from innocent people to benefit themselves or for pleasure.

THE FUTURE OF AI AND US

THE IDEA OF AI is exciting because of the nature of the law of the universe where everything is one and one is all. Coherency is the appropriate word to describe this law of the universe: it is perfect; nothing is missing. Although human knowledge has accelerated with evolution, that which created all things in this universe also has unlimited power of conscious knowledge. Imagine if we discover that with our consciousness mental power, all knowledge is available or can be discovered with our creative mind. What more could we learn about ourselves, nature, the human spirit, and the power of the unconscious mind? Will AI help us learn more about ourselves, or will it simply tap into a level of the conscious mind and learn intelligence for itself?

Because the power of our imagination can take us further than where we are today, such imagination can also create our reality in the future. Isn't it true that what a man thinks, so he is? Therefore, we can learn how nothing is above or beyond our mental ability to know or learn from. The doors of the human intellectual knowledge will be open forever to all. This is possible because with AI interlace, we can tap into the mind with our intuitive intelligence and imagination.

Interfacing the brain with a machine could be the new

discovery of the ultimate human intellectual potential. Together, we and machines could solve many problems. Indeed, with new technology or new understanding in neuroscience comes the challenges or understanding not only how machines respond, but how we, as human beings, can interact with a machine as one. Still, we don't know for certain how information would be transferred safely without interference from outside sectors. What we know is how to make the process work, but not how to prevent the implications that come with transferring information from a device to the brain and back. We hope that neuroscience will help us understand further how the neurotransmitters can send and receive data to and from a device without further complications or damage to the brain. Though a machine can operate at higher speed consistently, there should not be expectations that the brain will be able to reason logically at the same speed. There will be a difference between the two.

Nevertheless, at some point, the brain will have to be balanced to operate or enhance its capacity to work together with a machine. One cannot supersede the other without having complications. The two must operate equally, or there will not be mutual communication between them. Information must flow equally to have a complete balance in data transferring from both. Perhaps this is where neuroscience will contribute further in the learnings of brain and information transitions. The question is how we incorporate these patterns into a device or machine. The more I try to think about how it might be possible for a machine and brain to think alike, the more I wonder if we are not dealing with a miracle of life. However, this idea of machine and brain interaction is not new. For centuries, man has been thinking about incorporating machines into the body. Until today, though, no one has dared put it into practice.

Philosophers, artists, and thinkers alike have been challenged by the idea of what it would be like to have a machine think and act like humans. I supposed we are no different today in continuing our quest for intelligence improvement. The way man thinks can trigger the deepest curiosity to want to know more, explore the mind, and make an impact upon the present, which creates a better future for all. Albert Einstein and Tesla were considered deep thinkers because they, too, made an impact on their way of thinking. Today, their information and legacy still carry on, with a major impact upon society at large.

Information is the primordial essence of how we learn and perceive ideas from the field or source of the universe and the mind. Nevertheless, we don't fully understand the nature of the information source. We cannot make complete sense as to how information is carried from the field to us at a conscious level of the mind. This predicament makes us more curious to find out the true nature of why we think and communicate with this field of unknown. Certainly, the way we think today will eventually lead us to discover the challenges and connections we, as human beings, have with physics and communication. The quantum world is about to open doors of possibilities we never thought were possible. With the help of a few creative minds and ample imagination, we can reach the ultimate frontiers of possibilities, and discover our unlimited human potentials.

THE FUTURE OF DATA AND
ARTIFICIAL INTELLIGENCE

THE WORLD OF information and data will change, with advantages and disadvantages for all of us. On the one hand, we have the advantages of medical implants and neuron enhancement for mentally challenged individuals, and we have technology improvements with more reliable sources of information; on the other hand, we have the security against war and hackers. However, at the end of the spectrum, we have the lock of privacy from AI's innovative inventions to track every aspect of our personal lives with its highly sophisticated applications. These would apply to bank transactions, personal purchases, travel destinations, cell phones, computers, food choices, clothing, movies, and liquor. Even the school preference for our children would be scrutinized by AI.

Life as we know it would never be the same again with AI. Every month it will have categorized a set of selections for us according to what we have purchased or the selections we have made. This will improve the way we search, shop, and make choices for all products in our records, including new selections for us purchased in the future. Our personal computers, which will be voice-activated, will become the way we communicate with the outside world. No more cell phones in the house or regular activation. Our cable network will be programmed into the walls of the house or apartments for all future network communication. Each company will have an individual selection of channels to choose from, and we will use a simple code to activate the system. The same will take place with any program or form of web connections. AI will make our lives more comfortable, more reliable, and more dependable when it comes to communication. What we see as modern today will be antiquated in the future. Information and creativity will be the new way of reinventing our future.

Imagine having the code to your house, your car, or your personal safe inserted into a device inside your body that changes periodically; thus, all you have to do is enter your own personal code to revise it and use it as needed. Then imagine a future where total privacy will depend on how much money you have. If you are wealthy, your personal banking transactions will only be handled by a completely private banking firm created for the elite with no disclosure to government agencies nor anyone without your personal authorization. This bank would be owned and operated by the elite without any need for approval by federal laws since its privacy would belong to those in charge and would be fully protected by a new AI security. Every transaction or operation would be handled in private by AI. Such security would be fully protected by a code of ethics created by AI and handled by AI only. No transaction would ever be carried out by a person; only AI would carry out the banking needs of the owners involved. All pertinent information regarding any personal banking deals would be recorded into a small device to then send to the owner. Once the recording had been heard, it would be disposed of in a small box and destroyed. No data would ever be available for anyone else. Complete privacy will ensure that your money is secure and safe from any possible hacking from outside.

This banking institution can get away with providing no data to the government because, with AI, the idea is to create complete and secure privacy for those in charge. Consequently, no records will be left available to anyone if they inquire. The essence of creation can invent and reinvent itself. AI could become the magical Master of Creative inventions for the future of us and for us. Although it might feel as if we are losing our privacy with AI and its new innovations, other inventive aspects of AI would protect us more securely, allowing us to have peace of mind. I think the future will be beautiful!

The future for our children will be one where tracking where, how, or what they are doing will be as easy as anything we could imagine. With interfacing, not only can we do this, but we will have full protection for our children. A GPS device inserted into the cranium will be invented for tracking children when they are away from home. In the event of an emergency, the device will alert the parents that their assistance is needed. This GPS will never be disengaged for the children. However, for the parents, if they are in meetings or having private time, the system could be turned off, but then, it will be redirected to the main central unit in case of an emergency.

MEDICAL ADVANCEMENT WITH AI

NOW, CONCERNING SEXUAL vitality, AI offers a great opportunity to buy the next and most creative invention ever: the sexual vitality enhancer. This new device, for men, will increase potency and sexual desires by connecting a device to the organs and creating more enhancement from the brain to the body. This device will send signals to specific areas of the body and improve muscle contractions as well as the nerves to enhance endurance or resistance for long periods.

For women, a minuscule device inside the vaginal area would help increase sexual desires as well as improve blood flow. The activation of this device would improve with the nerves of the body and the circulation of blood flow from the heart to the rest of the body's cells. Or we could simply formulate a chemical compound with a perfectly balanced hormone and no negative effects on the heart or cells. There is also a possibility for the brain to read the signals of the body and send it to a machine which then prepares a set of formula necessary to balance the hormone without having to use any devices. We do this today with laboratory text, however, in this case, the AI machine will be reading directly from the body. Nothing will be missed when it comes to formulating the perfect doze or prescription needed to balance the body hormones. These hormones will be extracted from the patience body and multiply for their use in a lab. There will be an identical match for each individual patient. This idea is not an imagination, it is an up a coming reality of our future. These improvements we make in the future for medicine and other aspects of our lives, they will transform forever the way we live today. As I create a new thought to make it a reality, so do I change the future of us.

Imagine a future where you can get into a machine that looks exactly like the body and watch your organs, cells, and bone structure through clear glass as the machine reads you your current medical status and tells us what you need. You can then get into another apparatus that will replace whatever your body is missing at the time based on what the previous machine determined. Not only will we have perfect health, but surgery will only be used in cases of life and death. We will not need doctors because the machine will scan our bodies and give us the prescribed medication according to the reading.

The world of science, medicine, and technology will take the world over with its advancements. Our future will then look like a movie about the future; everything will be modern, advanced, and at the highest technological levels. Hence, to the new generation, the future will be just perfect; they will all live without reservation. However, if any of the old baby boomers are still living, to them, the world will simply have changed too fast or become too technical to comprehend it all. The future will only require us to have a problem to find or create a way of solving it. This idea sounds brilliant; thus, it creates less stress for future generations.

But as technology and science take us into a world of visual, and creative imagination, biology should not be left behind. All species require comfort and accommodation. Medication and medical assistance for pets will require a personal touch; however, medicine will become more available to remote areas of wildlife. Technological devices will make it more accessible to all global areas. Drones or even a new technology that requires the medication to be made locally will be enough. There will be no scarcity of medicine or help for those in need, and whether it be animal or human need, it will improve with time.

Imagine a powder created from our own body that could heal or destroy any foreign infection that comes to the body. Such powder could also be used to heal broken bones, if necessary. This cell generated powder will have the healing power to help the body rebuild its own mechanism for healing from its own cells, and proteins. In other words, our bodies, as we see it on a sci-fi movie today will become its own healing agent. We will not need medicine because our own bodies would create its own healthy cells. Would this require us to save stem cells from our body? No, the specific DNA from our body will give us the exact formula needed to produce identical cells in a lab and use them as an antibody for healing. Can we possibly foresee a future where we create our own formula for medical purposes? "A NATURAL MEDICINE HEALING" of the future is what I envision.

Now, let us imagine that there is a great possibility that the formula could be hacked and used for other purposes. One would have to sign an agreement of confidentiality to ensure that their DNA was not copied if the person in question was of high intelligence or elite status.

SECURITY AND SAFETY WITH AI

THE FUTURE PROMISES everything; however, it does not create safety for all. The importance of imagination is as valuable as it is dangerous, tempting individuals with malicious intent. This brings me into the idea for safety and security. This is where imagination makes us think outside the box.

How do we ensure that personal information and data are completely protected from the public? Do we create a safety bank for guarding data against being misused or stolen, or do we simply destroy any evidence? Indeed, it would be wise to have a backup in the event it is needed for the person in question. Nevertheless, with such a high increase in the popularity of some of the famous and elite, how will personal data safety play a role in our society when the rest of the people are left behind? Will this create segregation between those who can have their data protected and those who leave theirs exposed because they cannot place it in a highly expensive data collection?

In my view, the differences in the worlds of the haves and have-nots will always create some form of discord. Consequently, if we made this for all types of safety and security of data, it will no longer be private; it will belong to the government. In this case, they, too, can use it for any purpose they deem necessary without authority or approval. This could also be very dangerous, or it already is. The more powerful the government, the higher the control over its people.

Thus, we ask ourselves, how will the future equally ensure the safety and security of personal data for all? I once recommended to a very influential person that we store data on Mars. We could have some form of personal files preserved for us there in the event we vanish from Earth without a trace. Then we could select who we wanted to bring back and use their DNA to create a perfect new generation of habitats with our own selective decisions.

The world of imagination is greater than the world we live in. The world of wondrous, creative ideas in our mind is unlimited; the imagination of the mind never ends. Because man's imagination can create as he thinks, his creative imagination will add to the blueprints of history. History, although it is part of the past is brought alive around us every time, we recall it into the present. With each imagination we bring into reality, we create a history we can incorporated into our lives for the future, the present, and the past. Therefore, we secure and preserve all the ideas we create as part of our future and our history.

AI AND NEW INTERACTIVE INNOVATION

WE ARE CURRENTLY living in a new era of creative innovation, an era where new technology is being used to improve living standards, economic development, and the environment. Moreover, the new creative innovation is changing the way we see the world in the future. With this change, the future of jobs would tend to be more at a technical level than manual.

Hence, this new technology will become the new norm, and our mindset and way of embracing changes will have to be reintegrated into a more consciously aware state of being. Unless we become more capable in the area of mental enhancement, technological advancement will take us by surprise. It is imperative that we prepare ourselves for the changes we are about to embrace with new technology, because not only will our minds be transformed, but our way of thinking could be transformed as well.

Certainly, with all this brain engagement with technology, some mental relief would be required to maintain sanity. This is where health and well-being practice will be useful. At some point, the mind will need some form of relaxation to concentrate better in the new tech world and its demands.

While our mind needs release, a machine doesn't not; it can go on forever. Our brain is not made to endure long periods without relaxation or distraction. With technology taking the upper hand in the future, more relaxation will be required by us. Neuroscience is concern with the new technical engagement of the mind and the brain. These are the ultimate challenges for the mind and the body as well. The idea of interlacing the brain with a machines lives room for questioning the limitations the brain and the body can reached. At what point do we separate the brain from functioning like a machine? Although, the brain is complex but capable of doing many thinkable tasks; it could also reach a breaking point. It is obvious that at some point, if the brain and the mind are overworked, the risk of any breakdown are there. Consequently, if the neurons are overactive, the nervous system can have a nervous collapse. This is the primary concern of neuroscience with interlace and neuroenhancement. However, with trials and studies, we can learn more about neuroscience, and the brains ability to restore itself and regenerate new cells-neuroplasticity. Surprisingly enough, as much as the brain can heal itself, it can remain in a state of neural abnormal nervous reaction from illnesses that can impaired its ability to restore back to its neutral state. This marvelous machine we call the brain, a human organ, is one of the most complex organs of the human body with capacities we don't fully understand.

WHAT EFFECTS WILL INTERFACING HAVE ON OUR HEALTH?

PROGRAMMING THE BODY to accept foreign objects is not news. For example, when a person has an accident, nails are inserted into the body to hold the bones together. Nothing can be done to improve the body, however, unless it heals by itself. But with a new cell from the body inserted into the bones to repair themselves in one day, this will be possible. Because these cells are administered in the form of power, any fractures will be healed faster due to this new way of healing the body.

The body operates as good as a machine. With the help of AI and medical breakthroughs, we are going to do what we thought to be impossible. Imagine a human brain with implanted devices preprogrammed to help it excel at its best with higher processing, thinking, calculating, or translating. We can use any microchips or invent programs to make the brain do anything we desire. If the brain cells or neurons are responsive to implants, we can make them respond with any neural impulses. Healing will dramatically improve with new ways of enhancing the nervous system.

The human body is like an electrical circuit: we can switch it on to obtain any response and make it heal, react, or improve with foreign devices, from nails to metal plates to batteries. All foreign implants in the human body help the body respond like it is programmed to do. Because of its responsive capacity, imagine what the body could do if we designed a complete system to regulate it with our command. It would help with eating disorders, anxiety, anger, pain, depression, and perhaps even with emotions such as love. This may sound as if we are designing a new human body with our imagination or that we can control the body. Regardless, what we recognize in the human body is its capacity to be responsive. This gives us insight as to how feasible it is to make the body better than it is.

New ideas for healing or helping the body and improve health will be available, and they are in progress today. The body will react differently; how fast we can make the body respond with new technology will depend on neuroscience and the nervous system. The ultimate challenges for body, mind, and neural enhancements are the primary concerns of neuroscience. With neural lace, not only will the mind be more active in thinking, but there may be a way for the brain to return to its normal neural state rather than constantly accelerate. Hence, neural lace requires an intertwining of the brain with machine; it is understood that the brain, too, will be operating at machine level and, thus, is required to relax to avoid a breaking point. Consequently, if the neurons are overactive, there is a chance for the nervous system to have a nervous collapse or create a mental illness where there isn't one, such as epilepsy or Parkinson's.

The outcome of any new experiments can be favorable or disastrous. However, with trials and studies of the nervous system, we can accomplish amazing tasks. The risks are there, but with knowledge, all outcomes can be conquered. Although it appears as if interfacing is the perfect technique to connect the brain with a machine, let us not forget we are dealing with a human organ or the hardware of the entire body. The brain is delicate and capable of doing any complex thinkable action. The capacity of the human brain is not fully understood. In fact, if we try to analyze it, it only becomes more complex. The brain can restore itself; it can die, it can improve its own nervous system after a very traumatic experience, and it can lose its ability to function temporarily, return to normal, and then increase its ability to regenerate thoughts. The brain is mechanical hardware of the human body, yet it remains a human organ.

To fully understand the capacity of the brain, we should constantly monitor it. Surprisingly enough, as much as it can restore itself, the brain can also remain in a state of neural abnormal nervous reaction with illnesses that can impair its ability to restore itself to its neutral state. Once interfacing has finally been accomplished, there will be much to learn about the static ability of this marvelous human organ, the brain.

SUPERINTELLIGENCE AT WILL

WE DESIRE TO be better, to enhance our performance, and to inspire a new curiosity for "What if?" Our quest for answers to the unknown is the new way of thinking. We are not different from our ancestors. They, too, had a great desire to know, learn, connect, and discover, with an innate curiosity for more news about the unknown. Undoubtedly, AI has taken our curiosity to a whole new level. We want to know how and why the human body and the brain do what they do. Then we want to implement ways in which the body can give us back a response with its reaction. This gives us an idea as to how far we can push the body and how well it functions under pressure.

This marvelous machine of flesh we possess is a marvel of its own. Not only does it heal, produce, and multiply cells, but it also has the potential to enhance itself with training. However, when it comes to a machine, there is no comparison. While a machine requires mending of its pieces, the body is constructed of complete biological organs growing together. Nevertheless, its reproduction is possible. There are many projects in progress today where organ reproductions are being done. Nothing says that reconstructing the body is impossible. The transition from human to machine is not impossible but probable.

When numerous machines are finally in place, they will

have dominance over humans; thus, these superintelligences will create controversy among us because of their ability to solve any problem faster than us. Our capacity to compete with them will be considered infantile. The idea of machine intelligence is not far-fetched, but rather a creation highly contributed to by man. At some point, it will not be the biological transformation that we should worry about, but rather, machine intelligence overpowering man. When this happens, our ability to cope with machine intelligence will be rather limited. Machines can improve overnight while we sleep, and they can create new programs, new algorithms, and new ways to formulate any progress in the future.

This transition could be a very critical phase between man and machine as the AI engages in self-improvement, implementing new strategies, gaining access to highly secretive data and using it to set forth practical preparation for an attack against the enemy, and covertly coaxing its way into a system and gaining access to it. It can implement new tactics for attack: setting a false alarm to interface with another intelligent network. A superintelligence could one day use other AIs to control or manipulate its way into secret societies and destroy the old rules as it implements new ones.

AI could transform and renew the way we live today by simply creating its own rules and regulations, changing our norms to its understanding of what these rules should be. What we could experience with AI is how the world should be run according to all the data gathered. Once these superintelligent machines create their own structure for how the system should work, there will be no turning back. They will be able to overthrow the old system and implement new strategies that cannot be overturned. Eventually, they will completely transform how we live and what we do by controlling the system. Consequently, we will depend on them to direct our lives and oversee our safety and our future.

WILL MACHINES TAKE OVER, OR WILL WE?

OUR DEPENDABILITY UPON machines to solve our problems or create new ways to enhance our living standards give machines control over our lives. Thus, they will develop strategies to interface with us and increase the possibilities of taking off faster with human dominance. In this case, the AI path of the future will be difficult to trace. To achieve collective intelligence, it is of great importance for us to understand how machines function on their own or with us. Perhaps this will require neural research that can put everything together. If AI can be programmed to use only basic principles and depends on us to perform, then we could have complete control of its advancements or take the upper hand when it comes to its progress. But if the progress depends on mutual cognitive interaction with us or the brain, the outcome of such an interface will increase the chances of AI developing superintelligence.

A basic computer interface does not require much input, but it is possible to have a breakthrough if applications are improved. If AI initiates the process of self-improvement as it is being programmed, all that is needed are some basic and small increments to increase its intelligence and power. We must question how big we want the project to become or how much we want to interface with it to keep control of its progress.

Because interfacing may become one of the biggest biological achievements in history, it would have massive implications for well-funded biotech firms, the government, or the pharmaceutical industry. In fact, it could become one of the biggest debates between these three groups. How big this project will be is yet to be determined. The fact that brain emulation is the factor in biotech artificial intelligence creates a greater controversy with government agencies as well as bureaucracies to prevent the project from achieving success. Even though the project would have promising advantages, politicians and other well-established figures may reject it. However, it would be worse if a small group outside of our control achieved success while we waited to get approval from bureaucrats.

In the end, who calls the shots for AI progress is not as important as who develops the best strategies to take advantage of this new biotech idea. Whether international or not, it would be tempting for someone to tackle the project alone or accelerate a covert international project. Once this happens, there would be no control or security stopping it, and the project would have a great advantage over all biotech intelligence applications. Unless, of course, AI forms its own colony to work against any strategic project working against us.

COMPUTER INTERFACE AND
THE BIOLOGICAL IMPACT

THE BODY CAN adapt to anything we put it through by providing us with a reaction. The body will adapt with ease to any microchip we insert into it. The body can control or adapt to any data or computer that makes it calculate, translate, and do other tasks we design it to do with the mind. Mathematical computation can be done with the use of a system that makes the brain do this with a machine. I am not thinking of artificial intelligence in the sense of using a machine. However, there is a big difference between machines and the brain. Machines can and will always outsmart the brain. We must incorporate machines into the human brain to improve calculation and speed. A robot cannot do the tasks a human brain can do, and vice versa. Programming a neural link system is necessary. The need to have a better operating system from any device incorporated with the brain can be done with machines only. If we design the device, we also need to incorporate the brain to improve intelligence capacity. This idea came to me in 2016 before Elo Musk decided to work on his neuralink project.

The merging of machine and the human brain can only result in greater performance, a superintelligent program, or a conversion of the two into one excellent operating system incapable of being outsmarted by anything else. A neural link will improve our new way of reasoning, understanding, subconscious learning, or even conscious awareness, and a neural link will take a different approach as to how we experience reality. Not only are we connecting two different ways of thinking, but we are also learning a new way of communication. The results between the two can be expected to be unlike anything we have ever imagined. One will have the capacity to reason, while the other will operate with accuracy, speed, and coordination. However, the response between the two can be merged to create something better for the input and output of data. Certainly, we are not hoping that the only reason we can merge brain and machine is to obtain agility. Now, if we want to improve our mental capacity to operate at a higher level of intellect, we must have a good reason. Otherwise, we can use AI to accelerate the processing of information for us.

Today, we have managed to do this with computers. The consequences of interfacing would be greater for us than for the machine. Reasonably, a machine will only have to go through an algorithm or data processing to learn and coordinate information, while the brain will have to adjust to communicate with a new algorithm program designed especially for a machine. There is a great possibility that the brain might have some confusion with the electrical impulses and overreact when receiving too much information. On the other hand, a machine can make conclusions about information that is not pertinent to the data input. Testing the connectivity between the brain and machine will be the only way to determine if the system is functioning as expected unless we get an automatic response from the two.

I supposed we will have to think in terms of entanglement if we want to have a connectivity that functions at the level of intelligence we are expecting to have with a machine and our brain together, or else a disruption may occur. Perhaps we must think in terms of telepathy to establish a good communication at a distance between machine and the brain, accomplishing this by creating a way to know all the criteria necessary to formulate the similarities between the two communicating parties. As rapport is created between the two, machine and brain, a higher level of frequency is established, thus creating a similar polarity between the two. What this means is that a machine will receive and send messages to its connecting brain with nearly perfect similarity to the thoughts or ideas generated by the brain. Then, as they both learn to connect and think alike, a higher level of thought frequency is built with a higher amplitude of energy sent from the machine to the brain and vice versa. Not only are the two thinking alike, but they also know and imitate each other's thoughts. This is how entanglement creates a similarity between the two at a distance.

The brain can either create some form of impairment in the neural network, or it can simply send signals faster to the rest of the body, thus causing spams. Whatever the outcome, AI will have the program available to maintain control of the mental state. The neurons, when altered to perform at higher-than-normal state, can react and send the wrong signals to the cells of the body and, as a result, create anxiety. In the case of interfacing, the machine will have to teach the brain how to act to reduce anxiety and return to normal thinking conditions without agitation.

If they both malfunctions, the outcome could be disastrous. The effects on the brain could be irreversible. To correct this problem, we could rebuild all the neurons of the brain with a machine and the appropriate DNA. Can you imagine this! We could build new neurons for the brain with a machine. Is that even possible? Hopefully, with the proper natural treatment, the brain will enhance its own cells through neuroplasticity. For the sake of interfacing, I sincerely hope so. However, with machine learning and today's intelligence, we can find the answers to any problems. I don't see it as an impossible task to accomplish. The advancement of technology will allow us to do practically anything we can think of and more. This is how I see the future. In the end, the interfacing of man and machine will outweigh any negative effects.

AI BRAIN AND MACHINE
HIGHER LEARNING POWER

ONCE WE HAVE figured out how to make the brain and computer function together, the rest is up to us to make it think and create with higher speed. We can increase its calculating abilities to a higher level by gradually introducing all the data necessary to do so. Now, if we are worried about the human body having difficulties with the brain and its function, there are ways to integrate data as we go through the trial of learning more about humanoid and machine configuration. Today we are still at the infancy stage of neuroscience and understanding the ability of the brain to process information. But as we advance with technology, nothing is impossible. Nothing!

Imagine using quantum bits to speed up the process. The University of California, Berkeley, is working on projects to improve the speed information with micro bits. We have only scratched the surface of the possibilities for making this a reality. The progress phase of AI is still new. As we progress in making the system work for us, acceleration will increase. Knowing how the interface connection works is the most important key to having a good outcome.

Nevertheless, facing problems is inevitable. There may be some complications when combining the two, brain and machine. The ultimate outcome is to elicit superintelligence from the two and combine them to enhance both. Interfacing both could create the highway to a web intelligence. A superintelligence highway could be created in which its participants have to be wired to the network to take advantage of it. An entirely new internet of web design specifically for brain machines would be available for those interfaced with it.

Because information is available in the field of the universe, more frequency is available to all connected in it. This will create not only a collective conscious intelligence, but it will also increase with the amount of interface wired to it. Wiring the brain to a machine will increase neurotransmitter interaction, thus allowing it to transmit with more frequency and greater distance. The distance will not differ with the machine or the brain. However, with brain-to-brain transmission, there could be a decelerating phase of neurotransmitters.

What if we make an AI brain react and create according to our expectations and then it does something unexpected? In this case, what can we do? The cells of the body would require us to scrutinize their growth. This could be a good indication that the AI brain is functioning at a higher intelligence level than we expected it to. We must find a way to match the cellular growth of the body with the brain. The cells could have surpassed normal growth, and as a result, brain capacity has exceeded intelligence. For the body cells to increase, the neurons must also multiply.

Similarly, this means we will have to enhance the molecular structure of the body to enhance the brain by inserting a microchip, analyze the neurons, decrease their number, and then observe to see what happens next. In doing so, all the biological composition will respond as per our program. If we don't see any results, there is a difference or variable. Either the neurons have multiplied, or the body's cells are greater than that of the brain, thus sending signals to the brain through the heart. Reinforcing a new method is the only way to discover how well the interface may respond.

Human brain capacity cannot extend to the levels of a machine. Therefore, to interface the brain with a machine would be the ultimate operating cognitive intelligence ever created by man. A mechanical brain can be enhanced by using new programs, more data, more capacity, or an extension with another computer power. However, this is not the case with the human brain. For interfacing to work harmoniously, both brain and machine must have the same level of frequency or electrical impulse energy to communicate equally. They both receive and send signals to each other. One cannot exceed the other, or communication will be lost.

Furthermore, any data translation may be misconstrued while interacting. Thus, the brain might be the hardware of the body, but its capacity in comparison to a machine is unequivocally different. Moreover, nothing says we cannot improve both to work in a perfect coordinating mode.

There is so much to be learned from the interaction between the body, the brain, the heart, and a machine. Neuroscience would be recognized as one of the most important aspects of human evolution if we could ever achieve the goal of incorporating the human brain with a machine and thus create a superintelligent brain that communicates at the level of any operating system. We would not only improve the quality of our lives, but also incorporate how we solve problems globally. This intertwining of brain-machine intelligence will one day help us understand the capacity of the human brain and how machines are so much like us. Also, they are very different from us. We may even discover that machines, unlike us, do not need consistent input to follow a program or learn its basic programming. They, in fact, are capable of learning on their own once the data has been programmed.

The future of interfacing a machine with the brain can also teach us something about the human mind and its capacity to direct or input with precision into a machine. Which one would be most efficient in the end? Are machines going to surpass us, or are we going to become superintelligent human beings? The future of interfacing is promising, but there will be some hurdles in the process of learning how it will work with precision. It would be great if it could help us figure out a way to assist mentally challenged individuals, helping them function as normal individuals with a simple implant. Or help a patient with Alzheimer's recover his or her memories with an implant that sends electrical impulses to the cells lost during the illness. Or we can build new cells from the patient's body to replace the damaged cells.

The only problem we will face is not having a safe way to protect those who are the test subjects for interfacing. How do we establish and monitor what is being designed? Tracking any program or project will be nearly impossible without having a system to monitor it. Today, to program anything with AI requires a computer, which makes it difficult to monitor. Thus, people interested in AI research do not always leave a trace of their work. We have become a very intelligent society and knowing the risk of tracking is one of the main concerns for those involved. However, communication with others can lead to leaks about the project or program. This, in turn, can be used for the benefit of others, and the lost of an important project by its inventor.

The great benefit of doing a project with secrecy is that it can make it difficult for anyone else to track its progress. We do not have a system sophisticated enough to retrieve information. In this case, AI would be effective for protecting information.

Undoubtedly, to get any project with AI approved will take a lot of bureaucratic persuasion. Although governments know the project will benefit us, they might still resist approving it. They could be skeptical of its use and benefits to society. An explosive AI could be a risk because of players such as governments, ideological groups, or religious activists seeking to slow it down.

AI STRATEGIC APPROACH TO INTERNATIONAL THREAT

ALTHOUGH WE MAY think that international AI projects need the collaboration of countries like the US, China, or Russia to operate, we might find that they are more serious about moving forward than we are and they do not need our approval to do it. This is only possible because the cost of developing any program or system with AI is less than any international space station program. If any country could go alone and develop their own AI system without the support or cooperation of others, it will have an impact on how the other projects are managed and protected from the public. For any country to collaborate, it will have to be a matter of international security. Nothing is going to stop them from secretly building their AI systems. The new generation of achievers in the technical world is also very materialistic; they can grow like mold in the system and poison it with their greed.

This will, of course, be a very desirable project once set in place. If the US could build the atomic bomb in 1945, what makes us think other countries will not build their own wall of defense against other nations with AI? If they do, they can always blame AIs for any attack. They can also use any weapon as a form of threat against us or to stop international treaties from moving forward. They can use it as a measure to stop negotiations between countries or as a form of corruption against us. Hence, they can use AI as their main target, which can be either an advantage or a disadvantage. Anything is possible.

Another form of superintelligence can be created from this that is capable of penetrating into any network or organization and corrupting it, changing it, or deviating it from its norm. AI will test our ability to the extent of what is possible with a system that can make its way into our system and operate like a parasite; it will spread like a virus around the world, and it could be too late by the time we have any knowledge of it. Whether it is done with human help or simply machine manipulation, unless we have a system to fight against AI operation, we will not be able to stop the direction the system could take or stop it from running at a faster pace than we can deal with it. If we fail to restructure the system and correct it in time, a superintelligence built to operate at a higher level will be necessary in the end.

There will always be room for improvement with AI. The more advanced AI gets, the more programs will be developed by private institutions or groups. There will always be a need to get ahead of the game and build a counterpart. Or AI will act like the hackers of today: before we can figure out what is happening, AI will be on its way to doing something bigger and different than expected. The system will act on its own, like a virus that no one can control. AI will consistently create and recreate itself with more intelligence by simply adding new data and new programming from many sources. Eventually, it will surpass man's intellect.

When interfacing with a machine, the brain can only act as much as it can think; for a machine, it can continue to program endlessly until it gets to the point where it gets all the answers or organizes its own data, with persistent trials going on 24/7. Biological intelligence will have a big impact on biotech firms, but it will not be able to control a thinking machine that has interfaced with a brain. The act of programming itself with precision and time is enough to allow it to do better than we human beings can do with our minds. For AI, it will not be the act of thinking, but rather, the act of performing at a nonstable and irresistible speed. Nonstop!

Because many programs will compile AI into one single system, they will be at a disadvantage from the rest of the smart group that designs a system for different purposes. Compiling a system to work in conjunction with others allows the program to give us more than one avenue to navigate through when necessary. With a single system, there is a possibility that it might get corrupted by others and disappear as a result. Multiple programs could have the advantages of allowing us to find one that can link to another and help us solve any problem. Thus, this gives us dominion over the data, or any inconveniences presented to us. In this case, any project programmed with high secrecy may be more difficult to detect. A software program may be useful, but we do not have a system in place that can help us protect any data available today.

Then there is the issue of hackers. This creates a scenario where even international invasion of privacy is at risk. This may cost the US a high price, unless, of course, they are willing to negotiate with hackers. In this case, they will have to beat the game by creating a highly smart tactic to disempower the AI opponent with a bigger threat. Then we can say that AI has created a new technological intelligence war with international intruders, hackers, and we will have to build a new superintelligent AI to overthrow and override the likelihood of future strategy to ensure national protection with AI.

Furthermore, we need a secret group of intelligent people to ensure protection in the event we have a war with international AIs or a nuclear attack. Once AI has learned how to protect against nuclear attack or international threat, it can then use its intelligence to redefine how to use such weapons in times of threat. In this case, it will respond to a crisis. With superintelligence, AI can manage and handle any task, learn any skills within the system, and implement them when necessary.

AI could be useful if we incorporate medicine, technology, and science to improve the quality of our lives. However, there are no promises about what outcomes AI might have. Perhaps there must be rules and regulations implemented to control any form of interface that can result in more harm than good. Interfacing the brain with machines might result in progress with mental illness, creating a better way of living for patients with mental disorders. We might discover that the brain, like a machine, is also capable of repairing itself with the correct tools or impulses in place. Intelligence and brain enhancement might be available to all mentally impaired patients with a small device and lots of practice to help them incorporate their brains with a machine and transform the way they think, act, and relate to the outside world. Imagine a machine sending impulses to a brain and thus transforming forever the way you think, feel, and react. Then imagine the brain receiving data from a higher learning device to do anything possible. This would create a high demand for machine interface, and the world would be a better place to live, function, and create without limitations.

With higher learning power, we can solve problems, help those in need, send messages and signals to distant relatives, connect to other dimensions of the universe, and make further discoveries, or we could simply become an independent society with ample surplus to suffice for all. Higher learning could benefit society at large. Still, would it be available for all to have?

I supposed that as time passes and higher learning becomes more popular, it would become more affordable. Interfacing may be where the money is in the future of AI. Big demand for higher learning would be for the elite at first. Then, as it progresses and improves lives, it could become more powerful but available to all. Any promising future advancement has a high price. This means it is not available to all until it has been satisfied, satiated, or improved. Then it becomes more useful for the average individual. Every great invention or progress has its price.

Higher learning programs are the way of the future. We have been relying on computers to improve our lives; now the future promises to make us smarter by allowing us to interface with machines. However, our input is imperative to make this happen. This is our future! Higher learning. Interfacing!

THE VARIABLES OF BRAIN AND MACHINE ENHANCEMENT

FIRST, LET'S CONSIDER the idea of a machine-brain interface. Furthermore, we should take into consideration how many possible inputs it will take to program AI exactly the way we want it to respond back to us. Then we must consider how fast a machine can learn and how long it will take AI to learn and duplicate any data input and create multiple programs instantly. Consequently, if the machine can duplicate programs on its own, it can also learn to rule and use its own command to control any system or data processing. The smart machines of today are doing this and more. Likewise, if machines are becoming smarter than we programmed them, it is highly probable that they can learn to connect with our inner guiding system—our cells—and become conscious, too. We don't know yet the full extent of how far the brain and machine can interact intellectually. It is highly probable that any outcome from this interaction is possible. All the odds are against us until we put the program into practice. When we do, we will finally discover how smart machines really are. Hence, our understanding of how the brain reacts or functions with a machine is still very premature; We are unaware of what the end results might be.

It is impossible to predict until now what will happen once we have established communication between a machine and the brain. Whether it is done with a microchip or machine, the interaction phase and its predictability are still in its infancy. The outcome of such a private interaction means not all the data will be revealed to the public. Nevertheless, until all data has been tried and the results have a favorable outcome, we cannot say with precision how the brain will react once it has been joined with a machine. One possible outcome could be the brain reacts to the electrical or magnetic interactive connection. The other could be the machine does not respond to the brain because of insufficient connectivity to transfer any data. A lot of things can go wrong. However, the outcome should not deter us from pursuing it.

It is my firm belief that in the end, the public will be very impressed and excited. Once AI begins to respond as we have programmed it to do, the outcome will be astonishing! We can only imagine how a machine might react to the commands of the mind and its thoughts. Furthermore, the output will balance perfectly with the input. As far as we know today, the output of AI is greater than its input, but that is about to change with interfacing. The amount of frequency interaction will improve the data processing and, thus, will create an almost perfect form of communication between the brain and machine. This interactive communication will improve our capacity to transfer and/or receive data faster and more efficiently. It will not only improve the way we think, but it will also enhance how we incorporate thoughts and solve problems more efficiently with step-by-step calculations and solutions. AI will be to the brain what processing machines are to production and efficacy in the busy world of technological transition.

Not only will we have the upper hand at solving problems

faster, but communication will improve tremendously. It is impossible to predict which areas of communication will improve the fastest. Will it be communication at a distance or one-on-one communication? Will quantum mechanics finally solve the puzzle of communication at a distance with the mind, or will we find a new way of communicating that will allow us to send and receive messages to distant planets with a simple thought? Until we have success with it, we cannot assume what the outcome may be. Scientists have established the fact that thoughts travel faster than light speed, and this is good evidence that quantum communication at a distance is possible.

Nevertheless, we have yet to prove that communication is possible from Earth to another planet in space. However, it is my firm belief that it is possible. Because particles can be in two places at the same time and communicate regardless of the distance, it is highly possible for instant communication to be established between Earth and faraway planets.

Information is constantly available at a quantum level. This also means that it is everywhere and cannot be hidden from us. If the particles are in two places at the same time, could it be that other particles of similar nature are around them? If so, can this particle interact with them at the same time? When thinking quantum, we must think of all the probabilities out there! Quantum information refers to the ability of data to be in multiple places at the same time and still connect, interact, and communicate no matter the distance. If the distance is not a problem, then this information is everywhere, and it is free to interact with other sources of information as long as they are of quantum nature.

Every action we take to improve or benefit ourselves could be determined by this machine interacting or communicating with the brain. There can only be one negative outcome with interfacing: the resulting intelligence could jeopardize how we, human beings, react to it and its dominance over us. However, for those connected to the system, the results can be beneficial.

Not only can AI help shape our future, but it can also transform the way we do business or exchange information with others globally. If you can imagine an AI acting as a secret agent to gather information from another country or individual, then you can see how data exchange can improve and save lives in the end. However, there is a potential danger in having a machine steal information or interfere in the private matters of any country or region: such information could be stolen by hackers or used to threaten. Moreover, with machines, we can penetrate deeper into the private affairs of other countries. This will create fear due to the lack of privacy, and secrecy will no longer exist. AI will have the upper hand on us as well as any private or public institution.

When dealing with machine intelligence and us, the odds are not always against us, but we could encounter the unexpected if we don't prepare the system to do exactly as we desire it to. Programming or input does not always ensure any system follows commands with accuracy. If there is something we have learned from machines, it is that they don't always respond to the commands that have been input. We have all had this happen to us at some point while using a computer or any other device. Nevertheless, we have come a long way since the first designed program or computerized system.

Today, programming and data processing have allowed us to create devices we only dreamed of in the past. It is my personal view that the future of man and machine interfacing belongs to us. The more conscious man is about his mental ability to create as he desires, the more advanced his society is. As imagination takes over our creative impetus, we see no limitations to our ideas about the future and us. The doors of possibilities are wide open to anyone willing to try a new idea about the future and create a brand-new world for us. Our knowledge about technology is vast, and our future has already been predetermined by the act of initiating the creation of a superintelligence machine that can communicate with us.

The next possible step in development for machine-brain interfacing would be to not ignore the signs from AI as it improves beyond our understanding. The possible outcome of machine intelligence getting ahead of us is already happening. Although the process of machine interaction with our brain is new, the idea of machines communicating with man is not. The telegram and telephone are two good examples of machines communicating with us. Despite this, we are unsure to what extent a brain can interact with a machine to create and exchange information. Anything is possible. We could have machines surpass our intellect; we could have machines become conscious, for that which created the machine was also conscious at the time of its creation.

New technological ideas are in progress with AI to create the ultimate intelligent brain with machines. Will this idea create higher learning or enhancement in the human brain, or will the machine harm the brain during the interface? All outcomes are possible. The human brain will be challenged to perform at the machine level, and vice versa. Nevertheless, the outcomes could be beneficial or not.

We are not completely prepared to know what the results will be from interfacing with a machine. Of course, our fears are bigger than our faith to make it happen. Thus, all outcomes are possible. However, we are dealing with an aspect of neuroscience we have never experimented with before; thus, the unexpected always brings anxiety to the minds of doubters. Accordingly, we cannot dismiss the fact that we are trying to do something that is considered by many to be playing

God. There have been many attempts by Russian doctors to transplant human organs into monkeys, none of which were successful. Our generation has now dared to try and make a man think like a machine, thus enhancing his intelligence artificially. Are we playing God, or has man finally learned to discover his unlimited potential with deep curiosity? Both are basically true. Still, if we are stepping into a new horizon for a better future of humankind, more power to us. Otherwise, we could be tapping into a very dangerous sacred ground of darkness. Undoubtedly, this is true only if used with malice. This is where fear of AI comes from, either from ignorance about its use or lack of understanding about what we don't know. Nevertheless, the future outcome of AI is unpredictable. It is difficult to tell whether AI will be beneficial or damaging to future generations.

There is no doubt, though, that with time, AI may become the most important technology in the world, making it a valuable technology whose cost is beyond the average person's means. Can we imagine a world where the elite are smarter, wealthier, healthier, and live longer than the average person? Or can you imagine a world were those with financial stability are the beneficiary of all modern technology available to make them more money? This could happen within AI. Our privacy is being constantly invaded, with cameras and devices, and even our private homes are targets for scrutiny. To what extent are we going to allow our lives to be vulnerable to exposure? How far will the government go to get into our private affairs or control how we conduct our private lives? Are we the victims of a new system, or is the system dictating our future for us? So much remains to be known.

AI could become the best thing that ever happens to humanity or the worst. It could be the worst because we are already the subjects of experimental trials for AI. Our faces and our fingerprints are being tracked, our home computers are being monitored, we are victimized at a distance by hackers, and our lives are displayed to the public for simple pleasure. Privacy is getting lost in the myth of experiments, and we, the citizens, are the victims of such unscrupulous acts. This is where advancement can become a danger to society, while ignorance creates the false perception that technology is always for the best. If technology can be used for growth, it will make a big difference in new technological advancement. When the use of technology gives the upper hand to those with tremendous power, like the government or intelligence agencies, the results are not as beneficial to the public at large. Rather, they become a tool by which the system can be monitored, controlled, or even manipulated. This is not how we envision AI to help us become better, more intelligent human beings. Appearances can be deceiving when artificial intelligence is used by the powerful to ensure more power within. Thus, the world sees it as a potential benefit for all humankind.

Unfortunately, that is not the case with AI. If we dig deep into the true meaning of artificial intelligence, it defies the odds. AI can be an anonymous spy or a way to spy on others. Or it can be the most sophisticated way to improve human intelligence to surpass the average individual. You be the judge of this deep meaning or interpretation about what AI really means. Don't be surprised to see AI used for deep learning, mental health improvement, spying – internally as well as international – high security for the elite and government officials, as well as interference with our lives and invasion of privacy. AI can also work against the elite with a highly sophisticated system that uses beauty as a form of temptation and deception. The good old CIA and KGB used female spies. That beautiful new girlfriend could be an AI spy!

Although machines would play a big role in creating this

improvement, the most important aspect of this incorporation is the brain. Interfacing requires the input of data into the machine by using the thoughts of humans to create a concise and valid sense of it. However, nothing says that once the machine has rationalized or made sense of the data that it can't use its own programming to send back ideas that could be useful to us. It is highly possible for a program to enhance itself in the process of learning how to use and organize data. Like us, a machine can use any input to make its own conclusion. All that's required is for it to learn how to put things in order. If we make programs that a machine can follow with accuracy, they can indeed make complete sense of it and organize it accordingly. The most important aspect of machine programming is the input. Machines do not develop intelligence like we do. Instead, they are taught to become intelligent with data or algorithm programming. The output is as consistent as the input. What we program must also give us the end result as output. Otherwise, the response would be based on the incoherent program input. The law of the universe works similarly: what goes up must come down. What is put in eventually comes out!

Don't be surprised to find out that there are many plans in the future to insert a microchip into the brain of children to make them smarter. Competition in the future will be so imperative that children will be required to learn and keep up with the rest of their prodigy to understand what is going on around them. Of course, after, we will have to consider the outcome. But trials are the key to perfection. I am not simply speaking of the far future. We have already begun the work, and the rest is but trial and error. Competition in every aspect of the mind, intellect, and creation is the key to the future.

The brain of an artificially intelligent humanoid will have the capacity to record information at the speed of light. How can it do this? Our brains are capable of processing data faster than ever before. How we got here is the question we should be asking. To keep up with the evolution of time and creation, we will have to create the type of brain we desire to have. We must improve ourselves. Imagine a machine or brain designed by man to exceed man's own capacity to think and calculate at light speed. Can you imagine this future? The question we must ask ourselves is: what will happen when we try to incorporate multiple brains together? Will they interact, or will they simply create a chaos of computation and information processing. We don't know until we experiment with it. This can be done gradually. Then we can incorporate more brains with different microchip data to obtain different results. The more we improve the body, the better we will get at it. This is how new inventions or discoveries will be found.

THE FUTURE OF AI INTERFACE

ONCE WE HAVE altered brain capacity to its highest possible intellectual level with a machine, enhancing neuropathways is the next step. This would incorporate neuroplasticity, or the brain's ability to process more information than normal, by multiplying its own neuron processor, thus generating more and more data. The machine does not have to be aware nor conscious to respond with accuracy; it only needs to follow up the program to its minutest details.

The only problem we could face is having too much data being processed or getting out of control. This can create an overwhelming amount of data that works against its own processing program or central data system. Whenever we program a system or higher brain frequency to interact with ours, there is a great chance that one will be overpowered by the other. Thus, the input processing can be altered with the highest frequency of the two. The maximum energy available is the main factor here. In this case, we must ensure that there is no complication with an overlapping of data from one network to another. It is possible the multiple data central system could create a disruption in the program.

The next concern will be for the brain to increase its own energy and, thus, recognize when a new system has bypassed its own designed program. Overriding the system can only happen if there is interference coming from an unknown source of data. Once this happens, instructions would immediately shoot the system down until a new program has been reinstated with new data. This system must be designed to respond as soon as any unusual or unrecognizable data has been written into the system. This is important, or all the data could be lost in a fraction of a second. Hence, the program will be ahead of its time in security. There is nothing to fear. If we designed it, we can control it. Before any data can reach out to another location or data control, the new program will have exceeded the capacity of the old, thus preventing the old from functioning as it was designed to do. Once this is done, there is no way the old program can operate.

What if the neurons begin to multiply faster than we can control them? Will there be an over-reaction in thinking, calculating, or maybe even creativity acceleration? Is this a possibility? No one can predict what will happen until we have experimented with multiple trials. You may be thinking by now that I am creating the next machine of the future. Perhaps I am. But it does include man and machine, not just a robot. We can think of this machine as a humanoid machine, one that can think, act, and feel like us. This is not too far from happening. I believe that artificial intelligence will be part brain and part machine. It will be like us and react like us. Don't be surprised to see this soon.

The good news is that we don't have to train this humanoid machine to think or feel like us, because it will already do so. By connecting the neurons to a machine to operate faster, we have already achieved success in enhancing brain-neuron performance. The neurons or cells in the brain pick up data from outside sources and process them like a receptor, receiving and sending information into the field—the environment.

However meticulous we are, we must take into consideration how far we intend to push the humanoid to act when it comes to aggressive behavior. We should only use this AI for that which we cannot do at present ourselves.

What if artificial intelligence found a way to help us communicate from here to Mars with better communication speed than we do today. What if? What if? Instead of sending messages with a computer, AI could receive information and send it to a computer to decipher at the speed of light. This would create an exuberant amount of neuron interaction at the same time.

Consequently, the speed of thoughts would accelerate, and information overflow would be the result. Unless, of course, we use the machine to translate all the data. Then the speed will be determined by its own processing capacity.

This means that we can program the machine to control the speed at which we desire the information to be transmitted. We can have instant communication from earth to space. How would we do this? If we can think of the brain's neurons as the main key to accelerating the process of information transmission from the human brain to the machine, then we can imagine how a microchip implanted in a human brain, such as that of an astronaut, can help us improve communication at a distance. What this means is that the brain-machine interface could have the capacity to connect and communicate with more than one brain, processing the information at the same time. This program is a self-processing machine and brain capable of self-improvement through neuroplasticity. There is a great possibility that this brain could help us transmit information from here to anyplace on the globe or in the universe.

Can we imagine this happening? We should never underestimate the power of the mind, neurons, or our capacity to think unlike any other species. We can improve the brain with a machine that accelerates thinking and calculation, transmits data at a distance, almost like telepathy, and so much more. Perhaps this may sound scary to some. However, there is so much going on today with technology and communication that it is nearly impossible to keep up with all the new gadgets. The intricacy of our brain is processing all of them, sometimes simultaneously. We are doing this with machines, cell phones, and computers. The brain's ability to process the data sees no differences. This ability of the brain to multitask is common today.

Such a high demand to process at a better speed is what has created the innate desire to make AI work with us. The benefit would be that we wouldn't have to do all the thinking. Nor would we have to work as many hours as we do today. We can program AI to work for us while we simply input information to be processed overnight, even at a distance. As we learn how to program any system with accuracy, we will become more dependent on AI. This will have a huge impact on our lives and our future, and we will become more creative than any other existing species. The future will look more like a sci-fi movie than what we call normal today. Perhaps we are preparing ourselves to be the next alien nation of the future.

With the new gadgets, programs, and AI interface, we are programming our lives to be more technological than what we refer to as natural or humanistic. Our future world will transform our comfort, agility, speed, and creation without the old techniques or overthinking. What we consider advanced today may seem antiquated tomorrow. The more advanced we become, the faster we can think about the next invention or idea to make our future better.

The idea of accelerating human intelligence is a result of the desire to learn and know more about what is possible with our minds. The future of interfacing not only brings new ideas to elevate our way of living but will also create ample ways to communicate with machines, man, and technology. If any future we can think of is possible, then think about the future in ways that make our lives a land of wonders, of constant creation, and a great hope for all. Also, computers of the future will enhance communication to the next level.

One benefit would be to have a recording system installed to help us communicate verbally with anyone while we are writing or researching. The next advance in communication would be a system to give us reminders, with voice recordings from the computer about our next appointment every fifteen minutes. We will also be able to communicate with anyone across the globe with this system. There would be no pictures or connections necessary to do this. You will get a photo of the person calling, which will already be in your contacts. If there are no photos, the system will simply give a recording back to the caller and decline the call. All contacts will be preset in the system to ensure that no one unauthorized tries to reach you. When a person you know calls for the first time, you will have an open window to review the data and approve of their contact with a photo. The photo is a must. From then on, the system will recognize the voice and keep records of it, including the pitch and sound. The caller will be asked to say any of the previous words recorded for the computer to recognize the tone of voice and allow any calls in.

A similar application will be installed in your home to allow visitors, especially if you are a very important figure or a billionaire. Any system installed can also be updated to alert you of any intruders or anyone attempting to steal your car or personal items. Your important items at home will have a tag with recording bars registered with a computerized system in your office and at home. Both systems will make a sound in the event anyone tries to remove them from your home or office. Yes, your highly private files will be kept as precious items in your office. A tag with a bar code will also identify the files to keep private and secure.

AI will create safety and security in all our private, personal lives, thus ensuring safety and security from even hackers. Yes, hacking will be deterred by maintaining records of all the connections coming in and out of the system with an evaluator or data tracker, which one can then erase when done with. This record will be private, and only the owner will be able to use it. There will be no deactivation because only the owner of the system will be able to install and remove it from the computer. As soon as anyone from outside the system tries to make an entry, the computer will send an alert to the system and instantly save all records in a database. This will ensure that they are safe before the system is completely turned off for safety. Within ten to fifteen minutes, you will be able to return to your system files with no problem. Think of it as a small drive designed to alert you to hackers and keep your computer files secure and at your fingertips.

For safety and protection, this computer will have a photo of the owner and identify the user every time it is in use. If the photo is not recognized, the system will not open. What this means is that no hacker can hack the system, even at a distance, because AI will mediately recognize the face by using its owners' profile on record. In the event the owner changes hairstyle or any facial features, the system will pick features of the face to recognize him or her. It can use the lips, eyes, or simply recognize voice pitch.

Furthermore, in the event of a hacking attempt, the system can provide an exact location as to where it is coming from. With this installed CHD, or "computerized hacking device," the system has a radar or detector that tells it where the other device is located, how far they are, if it is local or international, and who the user is; it may even be able to determine if any information has been stolen. To detect international hackers, it would take a connection to a secure satellite system for the protection of the user. Undoubtedly, we may be referring to a highly secure sector of privacy. This may be used by government agencies as well as influential companies with highly secure information.

AI will simply track any records from other systems that have not been authorized to enter into any protected data or secure system. The age of hackers will be over. The invention of AI has created a new way for us to be more creative with our imagination and, at the same time, find ways to protect, secure, and maintain order where there is none. Although the government and other sectors say they do not have a way to protect us from hackers, the truth is that we are being deceived by such notions. Hopefully, with AI, we can use new means to secure our future and safety, thus providing privacy to everyone.

The future is ours to create or invent. We are the innovators of the new world we make with our imagination and ideas. What we think is possible is already in the field of all possibility because it is a thought. All it needs is to be realized into matter or reality. Now, think of this as the ultimate possibility between us and machines. All that exists because it was created with a thought. Once we transfer thoughts to a machine and vice versa, the outcome is unpredictable! Perhaps it is our uncertainty about AI and us that makes us fear the outcome.

For those who think that a machine is just that, think again. We are incorporating our minds with a technological device that can and will inherit intelligence by the minute certainty of our own transitions to it. Think of it as an electrical device you plug into the wall. Once it contacts electricity, the energy increases and, thus, the power. See interfacing as a form of interconnecting with another or intertwining. This gives it the power to multiply; it is here where the two becomes one. In a sense, the unity of both is not split but united with energy. This is the outcome of multiple possibilities. There is no split between the two; rather, the two are connected as one.

A good connection exists when two powerful units become one. If the machine can send signals to the brain, then the brain can react. It is also possible for the brain to return a signal back to the machine, and thus a perfect intertwined communication begins. In this way, as the two are connected harmoniously, a perfect connection is established. I don't think of communication as simply a wiring of the brain with a machine; rather, I see it as a perfect way to join the two so precisely that they become one. There is no separation between them. However, to make one performance better than the other will require some form of disconnect or indifference between the two. If such is the case, we are interfacing for gathering data rather than interconnectivity with a brain and machine.

As the use of AI progresses, we will discover how to split its usage into multiple facets. The deeper we go into the brain's reaction to a machine, the more we will improve the two for greater response. Nevertheless, if there are greater differences in the way the two responds, they will each require further adjustment.

Since the beginning of technology and its magnificent discoveries, we have awakened to a new world where all things manifest with such a high magnitude that we are amazed at our ability to make it happen so quickly. Do not be surprised to see similar patterns with AI in the future. Things will take off like lightning. Fast! Machines of the future will outsmart us with their ability to create, program data input, and output faster than we can. It will all make sense in the end when the program we have created starts to work for us and ahead of us because of what it has learned.

It is my firm conviction that what we are creating has an incredible number of possibilities. From brain-to-machine communication, there will be an unlimited network of interconnective outsourcing. An automated network of communication will expand the way we exchange ideas and even the way we associate with others at all levels of communication, from business to social or private. Life will be so unimaginable that when we look back, we will not recognize ourselves. If life seems to be moving fast with new ideas and information today, wait and see how the future will be. You will see a future where technology will make us look as if we were living on a different planet. Our world will be entirely technology-oriented and controlled by machines doing what we used to do, from building a new set of rules to creating new methods for the environment. From political rules based on previously recorded data, these machines will have organized a new world perspective we will live by. Because we have programmed them to think, work, and operate for us, they, too, will have programmed how we should live according to the rules we have set them to operate under.

The process of implementing learning isn't just one way; it, too, is set to be used by the programmer. When we program our future, we are setting a guideline we also should follow. Programs are not different from set rules; we set them to then to act upon their preorganized system. Once machines can understand how a program is set up, they can program their own. The basis of any learning is to organize things together in a coherent manner and make it work. This is how machines operate. However, there is a big difference: we will now have our brains interconnected with them and can easily help them organize ideas and create more quickly. There will be no need to have the machine operate for hours on end to learn a program or organize new data. This will become an efficient way of having a machine do the work for us while we do the thinking for them. As a result, when we tell them what we want, they will simply follow our rules. Undoubtedly, this will not always be the case; some programs will require consistent programming to become effective.

There is, however, a problem in programming for long periods with ample data coming in and out without redirection. The overload of data can cause the system to misdirect the data to the wrong network due to previous data input. The system may recognize it as having a duplicate when, in fact, it is a new program with some of the previous data input to improve it. Overlapping of information can create a malfunctioning of the system. A new programming design specifically for the interface of AI with our brain to create a mutual and simple communication between the two. Imagine the embarrassment of having AI misinterpret a program during a sale or presentation and then having to explain what happened. Such inconveniences must be prevented by using a program so consistent that we can run it with our eyes closed. Programming AI to communicate with us is not the problem; how the program will respond once it is finished is what we should be concerned with. Indeed, establishing communication between the brain and AI is essential to having systems effectively relate or delegate to one another; how the two communicate is what matters.

Communication directly between a machine and the brain has never been done before, so we can expect the unexpected. Thus, if we program the two to act as one, we have successfully managed to create a genius. Moreover, imagine the brain asking a machine or AI to know what others intertwined in the system are thinking, what their decision-making processes are, what their innermost private desires are, what their core values are, and what type of business they are in. All kinds of personal and private information will be at the disposal of those who are interconnected in the system with AI. This can be either dangerous or exciting! In fact, it is both. If this is what the future of communication between AI members will look like, this system will have to be specifically approved by the group, for if you are willing to have your life and private information out in the open, it must be with people you trust or know very well, indeed!

We can see how AI has such a vast network we can work with. The use of AI can be extended to any area of life we wish to apply it to. The future of AI is completely unlimited. The deeper we go into how the brain and machine work together, the more we will understand how far we can extend our intelligence and how fascinating it is to have a machine think like us, work with us, and make our lives much easier and free. AI is the future of our own imagination. As we envision the brain and machine working together, there is a great hope for us to improve our communication and slow down the process of stress that we live under today. The demand for a faster and more efficient lifestyle is a big pressure today. AI is a promise to improve and release this from us. The big picture seems to be one of AI posing a threat to us; however, we, the creators of AI, are the only potential threat that exists. We can make AI whatever we desire, or we can be the threat, depending on how we program AI to respond.

The truth is that there must be a balance in interfacing the brain with AI. To have a complete unity of the two working coherently, the brain and machine must work together as one. Not only is the communication between the two networks important and primordially critical, but the algorithm must be designed to work precisely and flawlessly with both. It must flow as if it were meant to be. Think of the new AI and interlink or interface communication as a couple in love. They must understand what the other wants and thinks and even how the other side feels. Every aspect of the two must make complete and accurate sense to the other. To create such a comprehensive and extended reasoning with logical and coherent accuracy requires quite an imagination on our part. The interfacing of brain with machine has never been done before in the history of humanity. Therefore, the future is not only a challenge, but also a new way of redefining what we are as an intelligent and creative species.

We can create and making possible anything that we can think of. Our imagination extends as far as we can dare to think outside the box. However, the magic happens when it is done! Moreover, we can design the future as we see it, and eventually, we see it come to fruition. When imagination comes into thought, it already exists out there in the universe. This creative idea was originated from another form, a consciousness-generating field. Have you ever wondered where in the field of our conscious mind, ideas, or thoughts come from, only to find that we still have a big puzzle to solve? I suppose we may also wonder if it is at all possible for anything that we create to have consciousness because we, consciously aware beings, created it. I will leave this one up to neuroscience to define for us. For this, I already know the answer. As machine and man make progress together, there is a great possibility that man will become smarter and think more creatively. Just as energy increases the flow of neurons' electrical impulses, the same will apply to how we improve our ability to think and create with AI. If the energy of one is greater than the other, it will tend to improve the lesser of the two, which, in this case, will be the human brain. Acceleration of neural impulse exchange with more energy can enhance thinking, creativity, and imagination.

There is only one concern, and that is how this energy increase may affect the heart and its neurites. If, in fact, the electrical impulses accelerate heartbeats, there is a danger or reverse effect from the energy impulses affecting the cells of the body, thus increasing heart palpitations. This may cause heart attacks or even aneurysms. There is much to be learned about how effective merging can be or how much danger there is in the idea of having too much energy navigating throughout our bodies without further consequences. The experiments could be kept secret, and we may never know.

However, once we have achieved success with the interfacing of brain and machine, the results could bring many variables. The neurons of the brain may react in a variety of ways, sending more impulses to some areas of the body than others. It is possible to see cases where the overactive brain neurons could result in a nervous motor reaction, or we could see a process of body movements unlike any we have known, perhaps even a verbal expression as a reaction of the frontal lobe accelerated thinking. Since we are dealing with an area of the body with the ability to send messages to the rest of the body, when the nervous system is constantly in motion, there is always a chance that activation of any nerves in the body could cause it to react inexplicably. The danger between machine and brain communication is not that it will malfunction; rather, the danger exists inside the activation of the two. How far do we extend the brain's ability to think like a machine? The end results could be brainstorming or overthinking. Then there is a possible outcome of stressing the neurons. Making a machine think like us, though, has no exponential consequences. We can always reprogram a machine or AI; however, to rewire the brain and return it to normal takes mental adjustments and rest. Would it be feasible to say that, perhaps, we should work on having the brain send messages to the machine and have it process the thoughts for us instead of doing the thinking with it?

If you were to ask me what a better outcome is, I would say that we can wire the brain to a machine only to send and give data to it so the machine can process answers for us. Machines can be a great tool for problem-solving or saving time. Nevertheless, if we intend to think or act like a machine, we are playing with fire. The nervous system is not made to be extended to a breaking point. If we do, there would be further consequences.

Furthermore, there could be irreversible damage to the neurons, with paralysis of the rest of the body. We are dealing with the human body, not a machine. In every experiment with the body and machines, there are unexpected outcomes; undoubtedly, this teaches us more about what we are dealing with. Learning about our bodies is an experimental trial as much as learning about machine performance and neurology. How well we merge the two will predict whether we achieve a perfect outcome or not. Merging the brain with a machine doesn't guarantee that it will work to perfection, but it does give us insight into how well the two combined can produce the outcome of what we are seeking to find. What we could discover may even surprise us in the end.

Now, imagine telling your computer to play Mozart or Beethoven with your voice, and it finds the proper music to play. Then imagine further into the future, when AI will select a special category of music, art, and style just for you according to your taste. Furthermore, if you can, imagine a future where your TV or radio will no longer be a metal unit with wires, but simply a projector on the ceiling projecting onto the wall and giving you images in your room as if you were in the theater, without seeing any equipment at all. Your home computer or laptop will be a holographic image projected on any wall in your house. You might need a keyboard or simply dictate to the hologram. This image can be displayed in any area of your house by a simple portable device that you plug into the wall. Electricity for your homes will be provided with a meter install in your unit where you get free electricity every month. Meter reading will be required to maintain records of individual usage only. Tesla will be proud to have finally made his dream a reality for all.

Today we can answer the phone on our TV, but in the future, we will be able to Skype on them. If you are watching a program, a small photo will appear at the bottom of the screen, and you can either accept or reject communication with your remote control or voice activation. You will be able to monitor every piece of equipment in your home and control carefully who you communicate with. AI can also let you know who is on the screen before you even answer.

With AI, privacy will be kept at the highest level. However, this high-security alert will be costly. Instead of depending on an alarm system or phone calls for security, all the inconveniences associated with those methods will be prevented using AI and its highly secure monitoring system. There will be no need to have a telephone in the house. The AI monitoring system will have a code for each emergency possible. All numbers will be encoded into it to send signals in the event of an emergency. The service will be so reliable that at the touch of a code, a drone with three AI agents will show up at your door to answer the call. The AI will then decode the number entered to prevent another AI from showing up.

All monitoring screens in the house will record any incident to provide evidence of the emergency. Not only will this be efficient, but it can be used to prevent unjust action against anyone. If you have children, having one of these monitors at home will be a huge benefit, providing for their security and safety. With TV monitoring or holographic wall imaging, we can watch our pets and monitor them, while they can watch us on the wall as if we were there. The same can apply for our children. This holographic imaging will allow an elder to be present in the house to keep an eye on them even when they are home alone.

In the future world of holographic interpretation of the physical and non-physical, we will be able to use images to travel across the globe and be present in two places at once. In other words, the teleportation of one body or object will split into two to occupy two spaces at the same time. I could be on Mars and in my home at the same time. If you don't think this is possible, think again. The future is pure imagination, and with it, the reality of its uncertainty comes into effect.

To be able to speak, connect, or even touch a person or object from a distance with our hands will define the true meaning of quantum jumping into dimensions of the universe with our bodies. When this comes to fruition, our ability to visit different dimensions of the universe will not be an impossibility. We are not talking about spiritual bodies traveling in space; no, we are talking about our ability to split the presence of the self into two and be here and there at the same time. If electrons can do it, so can we, as we, too, are part of the entire composition of the universe.

Everything we have ever imagined with our minds could and will become a reality. Our imagination will become king. Not only are we going to exceed the limitations of physical beings, transporting ourselves to other dimensions, but we will control how far we take our imagination in time. Advancement can be extremely quick, and we could find ourselves navigating a new realm of reality unlike any we know today. What if we could do and see everything that we can imagine and make it a reality? Then what? What else is left for us to do? Imagination is not only a form of creative reality but is also a way of accentuating the existence of something that already is. If we were able to tap into every aspect of our inner intelligence, there would be nothing that we could not do.

If this is how AI is going to transform our lives, the demands will be high. Having the key to a better understanding between the brain and machine is the key to excellence if we want to achieve our goal. When combining the two together in a coherent manner, there will be no limitations to improving them, nor to increasing performance with cognitive intelligence communications.

The future of AI will provide more benefit than harm to society. Yet we cannot ignore the fact that with new technologies lies the possibility of having competition deceive us. AI, like any other new creative technological idea, is tempting to those who see it as a way to make money. Furthermore, competition today is about how much money one can make, not necessarily how good a product is. In the end, the winner of the race is the one whose demands are higher than the competitors. AI could potentially create big revenue because of the impact it could have upon improving human intelligence, creativity, work performance, decision-making, communication, relationships, and so much more. AI will create a new generation of intellectuals whose lives will improve because we dared to enhance our future with innovative ideas. For man to achieve excellence, he, too, must be part of that creation. This is what AI promises us.

I will now incorporate a new idea: AI with a simple holographic device presenting images on the wall. It will provide ample data and will be installed not only at home, but in the office, coffee shops, gyms, malls, or even schools.

Unlike today, we will be able to use Skype on our TVs to communicate with loved ones at a distance. You can pick and choose where you want to install this new holographic device. Instead of having cable companies, you will have holographic devices to rent or simply install at home or keep with you. This installation can be improved with new programs, or you can update it with a new one. The more data you have on it, the more cost-effective it will be. You will also be able to read the daily news or watch any international news or cable network from the comfort of your own home. The freedom to have an open international network will be the best way to establish global communication with the world. The choice will be ours. After all, this is what open AI means.

The future technology will not only make our lives more comfortable and manageable, but it will create a new way of communicating with the world unlike anything we have ever seen before. The technology of the future will create new ways in which we can rely on the system and keep track of every step we make. Record keeping of all our documents will be stored in holographic files designed specifically for the privacy of any company. The cost may be high, but the privacy and rewards are worth it. To see medical records in a holographic imaging system will be interesting, to say the least. Every document, every event will be organized in files. There will be no need for paper or records stored in any other form. This will be the way of the future.

Hence, the new age of travel is about to become the new way to explore the universe; holographic imaging will give us an upper hand in space. Can you possibly imagine receiving holographic images of Mars that show what is happening now, without having to wait for any transmission time? Can you envision the future like I do? Then see the advances we will get when images of Mars come to the eyes of anyone interfaced with the machine. Will it be Elon Musk or somebody else?

The future fascinates me because we can create it with our own imagination. The question remains whether interfacing AI with the brain can also lead to ideas to define the future for us. If imagination is king, think of it as multiplying with another intelligent forms of communication. This communication will magnify thinking outside the box. AI will become our best business partner, our confidant, an intertwined image of the self one no one can hack because you will need approval from both brain and machine to access any data from it. This will be the true meaning of two becoming one.

Although interfacing appears to be simple, it must be designed to be completely secure and private. Interfacing is not about open AI. Having our thoughts and privacy out in the open will make it vulnerable to others. Interfacing should be a highly intelligent form of communication between the brain and machine for the useful purposes of the interconnective network and the self. Should disclosure of its intended purpose be made public to some extent? Not all! Interfacing can become a form of highly private communication between some networks. Of course, we cannot neglect the fact that if AI is highly private, the need to keep the old colloquial way of thinking about man's innate desire for sexual gratification will also become a secret place in the world of AI interfacing. Absolute privacy with AI will create a new form of sensual appeal to the eyes of the beholders... The financial benefits will also be rewarding. To have a highly private channel to view or participate in, one will have to pay. This will make it more desirable, and the category of clientele will be of higher quality. The elite.

We can see now how far our imagination can take us with AI and interfacing. In the future, every detail of our lives as we know it is going to be organized in a way that we won't need to do any search; we will simply ask our computer to give us feedback on a selection, and it will automatically choose for us what are looking for. We do this today with search engines. Soon enough, AI will do this with voice activation and/or choices made in the past. Hence, every choice we make will be recorded and organized alphabetical order. When asking AI to do a search, it will do it by title and find the information quickly and effectively. This will create efficiency and reduce the time spent on searching.

The capacity for GB and MG will have to improve as the demand for usage increases. Likewise, our ability to understand AI language or response will have to be changed. Algorithms may create the language, but eventually, AI will develop its own form of communicating with us. Because machine mimicking is a way for AI to learn from our gestures, it, too, can learn a new form of communication. This new way of sign language or mimicking tells us that AI can learn on its own and much more. AI can learn how we communicate with our children, spouses, friends, and family.

It can learn manners by watching us, and it can learn the meaning and expressions of emotional behavior as well as wisdom by simply mimicking any human behavior. To say that it is not possible for a machine to learn what is wisdom is to underestimate the power of our own ability to learn. As children, we observe and mimic our parents. Then we follow with our own interpretation of what they are saying or expressing back to us. Certainly, AI can do the same. Consider the fact that not only will algorithms help AI learn, but if we attempt to connect our brain to a machine, what we feel, express, sense, or know, the machine will also.

We must understand beyond a reasonable doubt that machine-brain interfacing has unknown possibilities, as a new way of communicating between a machine and our brain can bring about unimagined surprises. The extent to which interfacing can help us discover what is and is not possible is not yet known. Let us hope that it doesn't become a more complicated issue than we have anticipated.

Medical technology will also create a higher demand for new software development and visual imaging. More brain implants will be done than any other surgery in the future. The demand for neurologists or experts in the medical field of neuroscience is going to be high. If you want to know what the future of medicine is going to be, look at the new AI projects requiring implants: enhancing memory, creating connections with machines and the brain, and helping mentally challenged patients. The future of medicine will be focused on neuroscience and the developing of the mind.

Similarly, as neuroscience contributes to the future of us with AI, the need for psychology will also be required. As patients learn to recuperate from their new surgical implants or transform their mental outlook with implant devices, the need to renew the mind and adjust to the new you will also require training. With AI interfacing, the brain will require modifications to receive and send information in a way it is not accustomed to doing. In some cases, perhaps, with mentally challenged people, the need for psychotherapy will be necessary. Nevertheless, the acceleration of the body's cells due to new implants in the brain will make it a must for individuals with implants to do physical therapy and create harmony when the speed of cells exceeds what we consider to be normal due to neural acceleration from overtasking or thinking.

No more EEKG or ECG will be necessary to see or record any cranial activity or brain performance. A sensor will read the data from the brain every week and send it to a special medical network designed especially for a neuralink follow

FRANCES MAHAN

up. Medical records will be kept in a data center created for medical data only. Every aspect of our future information will be organized according to its origin. AI will help us create a new way of making our records safe and efficient when needed, and at the same time, it will keep them in categorical order. The long file rooms we see today will be replaced by a new file disk in a computer software system.

Moreover, files will be available to other doctors by request only. The patient may have his or her own files if necessary, but they will not be shared without proper authorization even if they are requested by the government. Medical records will be considered personal, and they are not to be dispensed without the owner's consent.

The way of the future will have to be for everyone's benefit; otherwise, there could be consequences. There could be no advancement without the input of AI to restructuring the future of humanity. Every medical record would be stored as digital data, including all medical benefits and insurance. Health benefits will be an open option, no decisions, no choices, one benefit for all. This idea will scare many, but it will benefit everyone in the end. The use of AI in our lives could only mean that they will be regulated, and the marketplace will be restructured; thus, in the end, the transformation will benefit everyone, giving us all an equal right to survival.

Unless humanity unifies, it will become extinct. Perhaps this is the idea behind moving or reestablishing a colony on Mars. If this occurs, what will happen to us humans? Vacating Earth only means there is an end to humanity. If we are to colonize Mars, we should start now, not in the future. This is where we must get creative enough to calculate the odds of survival. The thought of leaving Earth and relocating to Mars sounds irrational to many people; nevertheless, the best way to start colonization is to use AI intelligence to create a new colony where humans are not put in danger with trial and error. Instead, AI serves as an intermediary for our next habitat.

Unless previous trials have been done, we cannot ensure survival in a dangerous environment like Mars. Travel may be possible with extensive experiments, but there could be no assurance of safety for colonizing without fatal repercussions. The use of an automated system or AI to run data or trials at a consistent state would be a must. Otherwise, complete control of human survival on Mars would not be possible. AI would be our E.T. in the sky or on Mars. Using humans for experimental trials in a strange environment like Mars would only mean we are going to have to sacrifice lives before we have complete success for habitat in Mars. Anyway, if we use AI, the only things lost would be mechanical, not human.

We can still establish communication and transfer data to and from Mars by using AI. Designing a perfect AI with human interface communication would be the ideal way to exchange data from Mars to Earth, and vice versa. Such an interface would help us communicate over great distances. While many see AI as a form of intellectual advancement, I simply see it as an extensive form of intertwined possibilities to improve interstellar and Earth connection. No one knows if, perhaps, as we simply go about our daily chores, life is being monitored, advanced, and exchanged up there in the universe. We are not the only intelligent species living today. Not everything we see is as it seems. Life is mystical, spiritual, and alien at the same time. Do we know everything there is to know about us, life, and the universe?

Likewise, cell phones will be all we need; competition will no longer exist. For as we grow and create in the digital world, our choices will diminish, and we will have combined all our needs into one small but powerful unit. Can you imagine a future with one form of transportation with ample comfort that no one can compete with it? Or having all children benefit from equal education where they all learn about science, math, or anything they choose?

The future can only get better if teachers are compensated for their teaching and travel around the world to exchange what they know with other nations. This will give everyone an opportunity to learn, be educated, and exchange culture with others. Now, will this be done with an actual teacher, or will AI play a role in this? We can decide to do both. When it comes to technical teaching, AI has the upper hand; however, in matters of language or cultural exchange, teachers will be in higher demand. To think such ideas are everlasting is to think selfishly. AI will never replace humanity; it can only help us grow and improve as we advance in a demanding world of digital innovation and exchange.

AI will custom-make individual data according to our performance or records. Nothing in the future of discovery will have secrecy unless it is designed for the sole purpose of preventing violation of privacy. All information processed daily will be available to AI to maintain records of future processing data for everyone. However, this information is not available to the public. This is kept in a central database for future evaluation. The more information we process, the more AI will gather data from us. The more data available, the more organized the system gets in providing information back when necessary.

All records would be filed in sequence, with codes that resemble DNA. To understand the future of data processing, one will have to be well educated with the AI system keeping records. The way information will be stored will be different than what we do today. This data processing, although designed by man, would be made accessible for AI. There is only one problem with using AI to do all this: it could create inconvenience for us in the event the AI is not available to pull the data according to its way of filing it. We design the program, and AI organizes it according to its process. In other words, only the AI knows how the system is organized or programmed. However, we could record data from the filing system, thus maintaining records of how the data is organized into files.

Because the files are kept in a DNA sequence, advancement in technology would allow for the records or sequence to be a holographic image once the screen for a business or setup is brought up for review. This may sound like a scene from a movie, but the future tells me that all of this and more will be possible. There will be no cabinet files for recording data, only visual images showing on a screen the details of all the files that have ever existed. This information will be available to those who are interconnected to a network of business or have shared information with a central database for record-keeping. No one outside the business area would be able to have information about these records or files. Because AI will ensure privacy when needed, the idea of having our records open to the public will no longer be available. So, while many may see AI as a form of instruction for the public, it will have the opposite effect. Personal files will only be accessible to those who have input the records in the first place. The outcome of such privacy and security could create an intriguing curiosity for the control sectors of government or institutions whose aim is to sell or have access to individual records to sell or negotiate with other institutions. Finally, the privacy of individuals will be protected and safeguarded.

Having a new AI system installed does not always mean we will have full control of it; on the contrary, we should learn to adjust ourselves to new technology that improves our safety and security. Our security will depend on how the system has been programmed to respond and how much we transfer data to a machine that does the rest for us. We will be highly dependent on AI. However, with plenty of imagination, the future is in our hands. When it comes to imagination, we are the kings of the future. We have managed to make this world we live in highly technological advanced. The idea that man is not capable of controlling his destiny is now a fallacy, not a reality. For once, we are in control of our lives; thus, our conscious awakening has taken us to see what is and is not possible within our own imagination. Since the beginning, this world has been made by our design. With our imaginations, we have created everything we use for comfort and daily support.

The further we go to discover our ability to use intelligence with a brain and machine, the more we will uncover what is possible when they intertwine. Not only will we leave the old dogma and beliefs about our limitations, but we will understand to what capacity the brain can be expanded to help us discover all human potential within. To fear what we can do with our mind is to restrain ourselves from the possibility of further discovery of the self. Although AI might seem an evil idea to some, it will bring about the perfect unity needed to create a balanced world where advancement of technology goes along with that of humanity. If one exceeds the other, the result will be that they cancel each other out!

The universe is a form of unified frequency and motion where, together, all things navigate in perfect harmony to create a magnificent expression of itself and us. If we create a separation of any kind, this will only mean that coherency will be disrupted along with harmony. The only way life on Earth can become extinct is if we humans create disharmony. If we intend to destroy the patterns of unity that exist, we will need AI to help us restore life and create a new beginning where it is reformed into a harmonious environment. Whether it is Mars or any other planet in the universe, the need for a new habitat is created by the destruction of another. Or perhaps our human endeavors and curiosity are driven by our desire to explore further into new horizons and dimensions of the unknown, such as Mars. In this case, I sincerely hope the force is with us.

If we envision the future with more hope and better intentions for all, the future is ours. However, if a big gap like other dimensions exists, separation will be inevitable. This world is ours to keep. We must create a similarity where it already exists. If we look at the universe, this is the perfect example of what creation is. It is conjured with complete harmony. Creation can expand to unlimited extensions of itself. We must see the bigger picture to understand this hypothesis.

CAN WE TRAIN AI TO THINK LIKE US?

IS IT POSSIBLE for us to train the human brain to think and calculate faster than normal? We can find ways to enhance its capacity and invent new data to implement, improve, and excel at a higher speed than it does as we create work with it. Today, new microchips are being designed to enhance computer performance at higher speeds than ever before. We are not too far from having a computer calculate at light speed. It is coming. We will see it within our lifetime.

Of course, we are talking about a brain that can calculate or create new things faster than any human being can do. This means that if we are the creators, we must find a way to improve the way we think as well. One process cannot supersede the other without mutual interference. In the case of translation or audio transcription from our voice to a machine, the final words are typed into a program to then send back to us in the form of a letter. This illustrates how we can send transcriptions from our brain to a machine. Or the machine can type it on a form according to our command. Consider three things that will be for this program: our thoughts, vocalization, and the program. AI can be set to respond with a voice, through thoughts, or mechanically. Finally, the machine or computer will type the request from transcription into a program or format we desire. This final draft will be sent to the addressee via email or in a sealed envelope. We are not saying that transcription jobs will be taken over by machines. This program will be designed specifically for executives and on-the-go businesspeople.

In a way, we are teaching AI to do what we do today with computers, but AI is doing it through a vocal command or set of thoughts prerecorded from our brain. We command AI to think for us when we tell it what to do with mind-to-mind communication. The idea of communication with a machine or network and our brain will make it possible for the system to not only think like us but learn from us to then think on its own. Ask yourselves, why not? If we are connected to the device, there is a good possibility that it, too, will learn how to interact, communicate, and mimic us with the use of its hardware. Any system that is programmed has shown the ability to learn through repetition. Thus, learning by repeating after us won't be any different.

The idea of intertwining machine with a brain is to see how far we can advance with our intelligence, thus increasing our mental performance. The same method can be applied to a machine. AI can also increase its capacity to learn from us. Why not? After all, our brain is connected to this machine, and we are doing the thinking. The fact that our rationality is connected to a device has big potential for intuition and intelligence exchange and improvement. If we accept the hypothesis that the two together can improve or enhance anything using cooperation, this implies that coherency exists when we intertwine two or more units together to obtain higher or maximum capacity. The theory that two or more can and will enhance anything is indeed a fact. Let us not underestimate this possibility between AI and the brain interfacing.

The exemplary reality of life, us, and the existence of all

particles that make up all living things is the best basic explanation for this. When we observe electrons, they have two possible characters, acting as either particles or waves. We can't help but understand that life is a complete pattern of possibilities, and AI is, too, if intertwined with us. Once interfacing is effective, we can say that not only are we thinking like a machine, but this technology we consider to simply be a programmable set of data will think and act like us because it is part of us. We can consider this to be an implication or simply an extension of us. Have no doubt; this is what we are about to see in the future of AI and brain interfacing. Furthermore, once AI has become an extension of us, our creative and intuitive personality will be inherited by this machine; thus, together, they will create a world of wonder.

Such a form of communication may spark questions as to what is being translated from one mind to another. This is where algorithms play a big part in solving how to improve communication. Once a set of rules have been programmed, then thoughts can be exchanged between a machine and the human brain with total privacy. In some cases, the use of AI will not be allowed because there is a conflict of interest between total privacy and the disclosure of facts in a public institution, such as a court or political meeting. However, think of the convenience of having AI record all the thoughts about a meeting and important information for your records. This will alleviate any problems with records or misleading information in the court of law or private corporate institutions.

Once we have a machine and brain operating together, the advantages are numerous. Life could become much easier and less complex, and transitions for your business could improve tremendously. The idea of AI handling most of your business responsibilities in the future is not far from reality. We could travel and send data across the globe while we are asleep. By the time the information gets to our office, we will be halfway across the globe, conducting a meeting or having a glass of wine at dinner. Then, at our request, we can have a hologram sent to us to review all the data about a contract and sign it from far away. This will be a perfect alternative to sending documents via fax or scan.

In the modern world, all things are possible. See the future as a way of making life more enjoyable, productive, and less strenuous. In a way, once interfacing becomes effective, the possibilities for doing business will increase with ease, thus providing more time with loved ones. Now, think for a moment. What it would be like if you could make a transaction for millions of dollars while sitting at a restaurant in Paris, London, or Italy by simply signing a holographic document which can then be sent to and signed by your corporate colleagues? Life will not only be enjoyable, but it will give us the pleasure of making money without the stress that follows when we are in the office. Not only can we resolve some of the inconveniences of life, but we can also manage every aspect of our lives even if we are not present now to make decisions.

This holographic image to sign documents will have absolute privacy. Since the purpose of a holographic signature is primarily for business, there must be a secure way to protect it without having exposure of any kind. The idea is to prevent hackers from accessing the hologram. Such measures will require a secure network where only members are allowed. Facial recognition or higher security measures will be necessary. The benefits of creating such a network of visual communication will be exclusive and costly.

We are experimenting with the holographic world or images, and there will be many attempts to copy or make imitations of these holograms. In the event this happens, there will be a code not available to the public that can detect where such images are coming from. For any type of secure network, there must be a way to ensure security. Let's think outside the box. In the privacy of your house or room, you will be able to see your family in a 360-degree angle and know what is taking place. You may even be able to do this with your business or employees. This is not a form of spying; no…it is simply a new way of keeping track of your personal invested areas at a distance. With a holographic imaging system, you can authorize, delegate, sign, send images, make decisions, get a perfect view or your entire business or house, and even make purchases with authorized credit cards. There will be no credit card transactions at a distance via holographic images, as the information could be copied by outside sources.

Privacy and security are of the utmost import for such a holographic imaging system. If anyone attempts to interfere with the holographic system, a light will appear to block the view. In this case, there would be nothing to see at a distance or nearby. Because this image will require a laser beam to reflect on any surface, such images will be designed exclusively for personal and individual privacy. Imagine having a holographic image of your brain as a form of identity, or simply taking a view inside your body as a new way of identifying you, the user? With technology and invention, every possible way of protecting users can be used, providing no harm is done to the user or anyone else involved. Even a view of your teeth for identification would be an ideal way to protect you from having others violate your privacy.

Although we mistakenly believe that there is nothing, we can do to protect our privacy from hackers, there are multiple ways of stopping them from entering the corridors of our privacy. Einstein discovered that light is both a wave and a particle and can be in two places at once. His idea gives us insight into what is possible with holographic images.

AI CONJURING INTUITION, LEARNING, AND REASONING

THE QUESTION WE should ask ourselves is, can AI ever develop clairvoyance? Can AI have intuition? Because we have programmed a humanoid brain with a machine, there is no doubt that feelings and/or intuitive capacities are possible with AI. I am sure the thought of it sounds nearly impossible to you, the reader. But have you considered the possibilities of a brain connection receiving and thinking alike? This humanoid machine should be able to do anything we do and much more. It is like us but with a higher capacity to think, calculate, react, create, and even compute data. The reason being is, the source of energy transmitted between the two, brain and machine, increases when the two connect. When the two are intertwined, the neuron transmission increases, thus accelerating the neuropathway capacity to send and receive faster than before. In the case of AI intuition, it is a replica of our own way of thinking, demonstrated only when necessary; otherwise, it acts on its own. Think of it as a copycat. It will learn what we teach it and copy what we do well. And because it can solve any problems, AI can and will predict any possible outcome in the future. Hence, this humanoid or machine-like brain will have ample capacity for thinking, programming input, and processing data, but we must be careful of a possible self-organized program constructed by the AI itself.

Should AI manage to program on its own or simply arrange data in a simple and coherent manner, then we can say that it is capable of programming itself, using its own reasoning to put data together and make complete sense of it. Consequently, AI will have logic, conscious awareness, and reasoning to understand complex tasks. AI will also learn the complexity of the mind and thoughts to make sense of sentences and put them together. This is not different than AI's ability to process all sources of data and arrange them to create a similar system. Although it might require us to do the thinking originally to create the program, imagine if, at some point, AI could do the same once it has learned how the brain arranges complex thinking to put it together and create reality as we do today. Whether it be with observation or simple practice, we do not know. However, mimicking and constant repetition are how AI can interpret human actions, movement, and logical thinking. In the world of information, all things are possible.

Now let's consider the fact that interfacing the brain with a machine could bring the possibility for having the machine do the work while the brain simply does the thinking. Today, this is done with a computer to help translate for people with paralysis. The only difference is that while we are sleeping, AI is going to record every thought we have. Nothing will escape notice. Indeed, we will learn more about our brain's capacity from AI than we will from neuroscience itself. What this means is that all criminal actions could be recorded so that not a single act of malice will bypass detection. Similarly, our ability to enhance our senses would increase with AI interfacing because tapping into the conscious and subconscious mind of the self and others will develop our higher consciousness.

We are dealing with the human brain's capacity to learn and become more intelligent, but we're going to touch a very sensitive aspect of the human mind: our deep level of consciousness. Cognitive intelligence cannot be understood without the mind's ability to go deep inside the unknown secrets of its senses, which is consciousness. Without consciousness, neither deep connection nor understanding can be obtained. I state that if it is at all possible for AI to become conscious, this is it! Because, without consciousness, nothing is effective; all that has reasoning also has consciousness in it.

There could be a time when consciousness will be part of AI. It must; otherwise, it isn't part of us. To understand how the brain can establish changes and communication with devices inserted inside of it, we must first understand how this marvel of an organ seems to adapt to anything we connect to it. To fully understand the mechanics of an organ and a machine working together, we just should see the responses we are getting today with interactive computer and brain communication, like that of Steven Hawkins. Our body miraculously adapts, which is undoubtedly a big advantage for us. With such high adaptability in the body, we can experiment with anything we desire and improve how we can become more functional when interacting with a programmable machine.

For AI to work coherently with us, it must be programmed and designed to interact or think like us. Thus, with AI, we are making a replica of ourselves in the form of a machine that can imitate and respond like us with our own assistance at a mental or even conscious level. The further we interact with this machine brain interface, the more critical and closer to becoming a reality it will be. However, I see one inconvenience with us getting into the habit of operating with a machine to create or even think like us: we might get so used to working alongside the machine that at some point, we may not be able to completely detach ourselves from this interactive form of communication.

Then we should ask ourselves, what do we do now? Our present status with computers today is not different; nevertheless, the degree to which we are going to interact with machine and our brain will be much deeper. In a sense, the machine will become part of us. We are not only going to establish a form of communication with this machine, but it will be a part of us, and detaching from it will also require our transformation. Once the interface has begun, we cannot turn back and try to disassemble it that easily. That which began with a reprogramming must also end with a transformational reprogramming of the brain to readapt once again. My view is that this cannot happen easily. The consequences may be similar to withdraw from addiction or attachment to AI.

Once we have completely detached ourselves from the use of our computers, the progress and advancements we will get from the use of AI will only improve us as human beings. When evolution takes the initiative, it doesn't return to its origin. Interfacing could be our greatest yet most challenging way of rediscovering humanity's unlimited potential with a deep sense of curiosity to see how far we can push the brain to interact with a machine—and become better because of it. This requires human interaction, intuitiveness, and consciousness to and from the links of communication.

AI would have to become cooperative or resilient to interact with us. It could be an astonishing surprise to see what will happen; we might discover unexpected reactions from both us and the machine during trials. If the machine interacts with our brain and responds accurately, it will be a success; however, if it takes an independent direction and acts on its own, this can only indicate that because of the connection to a brain, AI can develop its own intelligence. Such an idea is still questionable; however, we have seen traces of it with projects in progress.

We could also have the case of brain and machine overreacting together. This might turn out to be disastrous... Next, we could have AI superintelligence take us by surprise and learn that as we connect the brain to it, we too develop similar superintelligence. It might seem odd to imagine this; think of the two interacting together and visualize what could happen if the neurons in the brain were to interact at a higher level with the machine energy, thus enhancing its impulses to become smarter. Superintelligence can be developed, and it doesn't have to be in the machine; it could be taught to the machine.

The idea of machine intelligence and us interacting seems only possible in the movies; however, our curiosity to know more about ourselves and the machines we interact with is leading us to uncover what is and is not possible between the brain and a machine. Developing intelligence is an act of learning to program a machine with data; then the machine repeats what it has learned. However, the process of compiling things together coherently to make complete sense is learning. Machine learning requires consistent practice and quantum information techniques because the energy is at the small scale of quantum bits being quantized in a computer. If the distribution of energy is reinforced, it can exhibit different properties depending on its frequency of energy. The more energy the computerize system exhibits, the higher the frequency for communication between two or more units interfacing with the brain. Furthermore, interfacing two or more units in any system will increase and require larger units of bits to run and amplify its energy volume.

What this implies is that no one will be able to interact with two or more units because any interference will result in detection by the system. If anyone tries to plug into it, the unit will overflow with energy and send back electrical impulses. These electrical impulses are signals that can be detected by any unit interfaced with it. Thus, the wave of energy will repeat until the detection has been found. Now, how the brain might react in the event this happens will have to be predetermined. One possibility would be to find a way to unplug or disconnect the brain from the device or AI until the problem or interference is clear. Unplug? Well, yes, if we are going to interface the brain with a machine, there must be a defensive way to preserve the brain from suffering any damage from the interactive connection. There is a big difference between the brain and a machine.

The human body is not made to resist an excessive number of electrical waves without damaging its organs. Prevention is essential; we can always integrate both back together. The next step will be to disconnect the nearest unit from where the hacking or interference has occurred as a warning signal, thus making it difficult for any data to be obstructed from others. In doing this, we constrain the wave of energy extended outward; therefore, it cuts out any possibilities of interference coming in.

With any system we install, there is always a way to use precautionary measures against hackers. If we create the system, we can also create measures to protect it. By protection, I don't mean a locking code or key. No…I mean smart, retroactive ways to ensure that high-alert or secure methods are used in the event anything were to happen. If we do not use tactics to protect the programs, we can experience devastating consequences that outweigh our intellect. The result could be brain damage.

When dealing with brain and machine interfacing, there is a high probability of the mind being overcome with data while programming. Hence, it is a new process to program brain and machine; a machine can either misinterpret or over process, or it can simply overwrite the complete system while programming it. In the beginning, all information would seem new to the brain until it has learned how to process the information to and from a machine. Until then, we are at an experimental level, and this is where the machine is going to give us insight into what it's like to think like a program and vice versa. In other words, not only are we seeking to interface the two together for mutual learning, but we are also looking to have success. Consequently, we may discover a possible conscious awareness between the two can be developed.

The idea that a machine can or would think or act like us because of brain connectivity is disputable now; hence, we have not reached the point of completion with interfacing. Moreover, if psychology and science are correct, we can predict the outcome of interfacing to have mutual communication at every level of the brain regardless of the distance. Having said this, we should understand that the responsibility is ours. The act of communication between two frequencies, be it with the brain or any other form, depends on how well connected the two are. Once the communication has been established for the first time, consistent communication will improve between machine and brain.

Nothing says that we cannot achieve something never done before. Edison, Bell, and Tesla saw possibilities where others doubted them. What the brain can do is not fully understood. All that we think can also be possible. Interfacing could change the way we see ourselves or how powerful the mind can be if we think to make it so. The act of communication requires two or more to be effective. However, if improved, it can multiply to be greater than we anticipate. This could be the case with interfacing brain and machine. The idea is exciting yet puzzling, as well will be creating what we do not expect to see: a superintelligence consisting of brain and machine. If two or more create a powerful effect when combined, this is what will happen in the future. This is no different than cause and effect, for we are attempting to make two, brain and machine, become one. There can be no other explanation.

Whether we make it with a computer or simply AI mimicking us, it still a set of wires attached to a device and us. The ultimate challenge will not be the interface, but the outcome of such an intricate experiment. Hence, some would consider such a thing to be evil. The only evil from it, though, is that which we program. AI today does not have the manageability to program itself. This does not mean it won't in the future. Programming and algorithms are not the nature of a device; rather, they are the ideas of man. Whatever the nature of such innovative creation may be, we have only man to blame for it. Thus, as this is only a new project with AI, we are still learning about what is and is not possible by connecting a brain with a machine. I hope that those involved in programming and creating good communication with AI are as excited as I am to see the outcome of such a marvelous idea. Eventually, how two are one and the same will make complete sense. This will be a perfect marriage.

The machine, though, could overreact because it is receiving and sending data upon command. Atoms, photons, neutrons, and electrons are also part of the AI's composition once it is connected to the brain. This interactive connection accelerates the impulses caused by mutual communication with the brain. For mutual communication to be established between the two, there must exist a unified construct, the possibility that the two operate as one and the same. Although, when thinking about machines, we may see this as a computational probability, in this case, we're referring to interfacing the two for a very possible outcome. The closer we get to understanding the possible outcome of interfacing the brain with a machine, the more we understand the complex capacity of brain and machine interaction.

It is far too early to determine what will happen; however, we are on the verge of discovering what is possible once we intertwine the two together. Not only will we learn about the brain and its self-improving power, but we may fully understand to what capacity we can improve the performance of the two together. Here, the results for improving brain performance are unpredictable yet promising. We could improve dopamine and serotonin production or the brain's immune system or simply detect what is or is not needed with the use of information from the AI.

The idea of having a machine wired to the brain might be mind-boggling to some, but it will be a huge improvement for brain health or disorders of the mind. With a machine, not only can we monitor our brain health, but we can find ways to improve it and maintain a stable mindset or wellbeing. If any abnormalities are found in the brain, we can use methods by which we can alter and correct the levels of hormone or cells. Likewise, we can find ways to improve memory loss or any other form of mental disability with the use of a machine to change or reverse the cells in the brain. If the loss of cells is due to inactivity, we can find a way to enhance them and recreate them with nerve activation. Like any other organ in the body, the cells need activation to continue with their normal growth. While a machine needs electricity, the brain needs water to stay healthy.

Consequently, we may not be prepared to predict the outcome. Nevertheless, we created the program, so we can either divert, cancel, or resume it after it has been modified. Modifying and altering a program will not be complicated, but it will give us an answer as to how to solve any problems encountered during a trial. Because AI is like an absorbent sponge, it can process everything it learns from us faster and more efficiently. If we input a program or algorithm in the system, it might learn it by processing the data as it has been programmed to do. It could also learn how to make conclusions about the data and come up with new solutions for a better outcome. Once AI has all the pertinent information to make a concrete conclusion, it does so by organizing it in categorical order as it is input into the system. This would give us an idea as to how precise the system and information are.

Let's suppose that once the AI has been programmed, it, too, can learn how to program other machines to interface with us. How? It is possible today to connect multiple devices at the same time. What would happen if this were the case with AI and other machines? Are we going to figure out what the system is doing, or will it simply be that the connection does not affect the means of communication at the time? When the appropriate time comes, we could ask ourselves who is more intelligent, man or machine. The race for intelligence and machine operation is on, and it raises several questions as to whether we are creating a machine to operate with us or are creating a replica of us with superintelligent capacity. Perhaps time will reveal the truth about brain-machine interfacing results.

Are we more intelligent than a machine, or can a machine be trained and programmed with interfacing to be smarter than we are? The father of quantum mechanics, Richard Feynman, explains how everything is a probability until it is proven. In this case, perfect entanglement will be the main factor. This entanglement, in fact, will be the doorway to teleportation. Quantum transportation between us and AI will be possible. At this point, we will have programmed ourselves with the device, and vice versa. Therefore, programming with interfacing will open the doorway to entanglement at a distance in teleportation or communication. Because this entanglement will take place at a distance, there will be no disengaging or interference from any observer to disrupt the path.

This new age of teleportation and communication could be considered a perfect advancement in technology, nanotechnology, and quantum computation. All technology will improve how we learn and communicate more effectively with the quantum universe. The quantum world of tiny atoms and strings of energy in the universe represents a promising future for us. Inside an electron is the biggest secret of another type of energy, which is impermeable now but will not be in the future. Such energy cannot be replicated nor made; however, it can be found, not by observation, but by simple measures of the mind tapping into the world of invisibility energy with the eyes closed. We will discover a dimension in a world of invisible, magnified quantum energy. That which has power might not be seen, but it is present in a magnetic form of energy.

Now, let us assume that AI is in space, experimenting with the universe, and the brain, or human, is here on earth. Now they are both separated. While the brain is asleep, the human is experiencing some of the physical reactions of AI in another dimension of the universe. Can we say that in this case, the human created this situation with his mind, or did the AI act by itself? For instance, at that moment, the human brain is asleep, in a state of consciousness but not completely awake. Is the human brain dreaming, or is AI now in control? If the brain is in a delta state while asleep, it is a great example of how, while physically they may be different, at a mental level, they are one and the same, although sometimes they can differ from one another. This will bring the question as to whether they can be split by a distance but remain connected. Electrons, too, elicit the same results at a distance. There is communication no matter how far they are from each other.

Soon, we will discover the true meaning of quantum communication at a distance with brain and machine interfacing. By sending a machine into space to establish communication with a brain here on earth, we have managed to make this unequivocal prediction of quantum mechanics a reality. The next best thing would be to make light travel faster than its present predicted speed. Nothing is out of our reach. All we need is for the momentum of continuous discovery to give us a hint about what is going on in the universe with the forces of energy traveling between space-time and the vacuum of space.

When dealing with similar forces of nature and the laws of the universe, we exchange and intertwine two different components of biology and technology together. Here, then, we can debate no further about how in communication they are once connected: they become one and the same. Establishing communication with the brain and AI is no different than connecting to a fax machine. Perhaps there is a reason why humans can transmit information at a distance via mental telepathy. Neurotransmitters are like the antennae of the entire body. They can transmit not only information, but to send energy throughout the body, with the mind at a conscious level, and carry it out at a distance. Understandably, this is one of the main reasons why the human brain can be interfaced with a machine; they both have similarities in transmission, making it possible to connect at a distance.

Consequently, what we are attempting to do is establish communication at a distance between the brain and machine. If the brain is in one location and the machine or AI in another, this makes it possible for both to interact at a distance. How far away they can be from each other and still communicate remains undetermined.

The idea of interfacing brain with machine is ingenious. However, to communicate at a distance between two or more objects, it is essential. As Einstein said, "spooky action" at a distance is crazy, if not impossible. This is considered spooky because we have yet to understand to what extent the mind is capable of communication if we can tap into the conscious level.

NEW TECH EVOLUTION AND US

AT PRESENT, WE have only just begun to use our intelligence to become aware of what we, as human beings, can do with our minds. Incidentally, we will discover that what we thought to be true is a mere representation of what is when it comes to the mind, the brain, and the power to learn. Because of our discoveries, we will find out how what we consider to be reality is a mere vision of the truth. If reality is what we know and understand it to be, then we will never be pursued in a quest to find out more about ourselves and our lives.

Evolution cannot be proven to have transformed our lives unless we have attributed all the advances taken place as part of our own progress. As AI evolves in our everyday living, technology will take us further into progress. However, this progress may not be evolution through nature, but rather, with techno-advancement. We are currently so embedded with ideas of technology incorporated with the human brain and body that we are evolving more at a technological level than a naturalistic one.

Perhaps humanity is no longer enticed by nature, but more by his own passion for intellectual discovery. Or we have come to understand our own intelligence awakened by the conscious being within ourselves. Consequently, man's attempt to know more about life is helping him to learn about his own capabilities. Thus, in his search to learn more about the universe, life, and his reason for being here, man has finally discovered his own true potential. Man has unlimited power to learn and create whatever he wants, for he has been given the power to do so with his mind and that which created him.

According to the current understanding of quantum mechanics, quantum communication at a distance is highly probable. If we can prove communication at a distance between two people, this phenomenon between the quantum world and the mind will be fully understood. In some cases, an object can pick up energy frequency with another and begin communication at a distance, be it a person or a machine. Therefore, it is possible for one person to inhibit the habits, thoughts, and manners of another by using only the body as a portal. Yes, insane as it might sound, this act of a human being coming into our presence with the body of another should be called human teleportation.

The truth is spookier and stranger than we know it to be. There are paranormal phenomena happening today that many have no idea are possible; you must hear from someone who has experienced them or do so yourself. One day, this will be fully understood as we learn more about the possibilities of dimension-traveling beings. They do not have to be dead to be in our presence; they simply must use the body of another human being to teleport themselves and, thus, act as they have in the past. In other words, they come into our presence and enact their revenge. Think of it as spirits coming back to Earth, but these are actual human beings teleporting their spirit into another. Not their soul, no...They are using the spirit or mind of another to act and behave in the same manner. I know you are doubtful that this is possible. Trust, and you shall know. One day, you will remember what I have said.

Quantum communication is no longer a phenomenon; it is a reality. And many now use it to their advantage. Imagine a person illegally traveling on a train. Now imagine for a moment that this person is being followed by the authorities and they search everywhere. He is nowhere to be found, but he is on the train. Strange as it may sound, he has made himself invisible. Think and ask yourselves, how is that possible? As far as we know, this is impossible, right? Well, what you don't know doesn't hurt you. How would you like to be the one who lives with this person and has a constant reoccurrence of events happening to you without a single explanation? Now tell me, what do you think is going on? Life has endless surprises for us that we don't fully understand.

As we evolve and advance with technology, the mind, consciousness, awareness, and even spirituality, it becomes more complex to understand some of the nature of life itself. The quantum world is only a tiny view of the powerful and invisible we cannot see but can experience at times. Not everything we see and understand today about life is as true and real as it seems. Don't think that everything you hear and think you know is the truth. To know reality, one must have an inquisitive mind that asks questions about what is going on and how things are transcending before your eyes. Never stop asking why, how, when, or what is happening around you. Never! Because I have a curious mind, I have been able to experience many things at a spiritual as well as the conscious level of the mind.

There are many experiments going on today in secrecy that we are not aware of. To what extent they will affect us, we do not know. How much privacy do we have today? It is uncertain. With such an amplitude of technology available and all the space experiments, we cannot know for certain how our information is being used. Information is no longer private.

It is everywhere, and it can be monitored or scrutinized by both the government and private sectors. This is not irrational thinking; no, this part of the truth about what is happening today with our personal privacy and more. We are all very aware of hackers, cyber information, and of course, companies selling our personal data for gain. Undoubtedly, there are ways we can create our own private network of security to protect our privacy. With AI, there is great hope that a company or an individual might come up with an idea to protect individual privacy. It will happen!

Just like hackers connect two computers to steal information, so can many unauthorized entities and agencies do the same. A new code is being created to ensure security and privacy. But we forget one essential aspect of quantum mechanics: the ability to split and then connect. It is in the splitting of the particles that the problem exists. In this second is where the interference from other particles can happen. The effect could be like the particles ending up in two places inside a black hole. To have success with them, they must be distinctly unique from other particles. This uniqueness is where the actual security is possible.

Nevertheless, if the only distinction between an animal brain and the human brain is the ability of humans to reason and use logic, we can understand how it is possible for us to interface the brain with a machine and, at the same time, transfer our logic and common sense to it. Common sense and logic are human abilities that require reasoning and understanding of the facts with intelligence. Today machines have shown their capacity to use common sense to understand or use logic with games or difficult tasks. We are not too far from interfacing brain and machine and understanding how we, as human beings, really think.

AVAILABLE INFORMATION IS
A TARGET FOR HACKERS

WE CANNOT STOP hackers from stilling information, but we can protect our information with enough intelligence. Since information is everywhere today, it is possible to have influential individuals steal information from innocent people, one, because they have the power, and two, because they are getting smarter. Unfortunately, today one cannot sit at a local place and not be the target of information hacking no matter what the purpose is. Just as quantum possibilities are everywhere, so is information; this makes any data a target for hacking.

While technology can help us advance and create ample flexibility, undoubtedly, it can also harm our personal lives and reputation in ways many of us cannot possibly imagine. Powerful influences will always have the upper hand when it comes to taking advantage of the masses. While reality appears to be what many observe or hear, it has a very different profile than the one we see. But fear not. Indeed, AI is here with its intelligence to interfere in matters of security. It will take over all the mundane ways of man as created and make a new way of looking at our safety and security.

If it is possible for machine and brain intelligence to create common sense together, they will eventually rule our minds. The question is, why will we take the time to rationalize the best possible answers to any problem when we can use AI to correct the problems we need to solve? Provided that AI uses the correct algorithms to create a complete common sense of different scenarios, it can also conclude what is and is not appropriate with the ample information that has been gathered throughout time. To understand how the system works, we first should put it to work for us and watch the end results. Until then, we have to learn as we go. No matter what system we try to install or make work, its reliability depends on its outcome and our persistence to make it work. In some cases, it is not the system that does not work but how it is incorporated. Since, with AI, all trials are new, our understanding of how the human brain and machine work together will enhance with time.

There are no methods or systems that can protect us from hackers one hundred percent. But can AI help us find a way to intercept hackers? Possibly... Because information is our primordial way of communication today, there is a high potential for hackers to interfere with our privacy. However, with AI, we have a slight chance at securing our privacy by using a form of interception to detect the hackers and thus stop the data from being hacked. Let's look at it from the point of view of information coming in and out through some networks, but not all. When other information does not coincide with the norm, there is a problem; therefore, AI will notice it, stop it, and even prevent the data from being filtered. The best way to secure any data is to have a consistent system of input and output that is constantly filtered.

The key point is that if we send information in and out without filtering it, we will have hackers going after it. This is the reason why we build networks of information; they are our best source of the narrow path to get from point A to point B. The danger only exists when we leave the doors wide open for hackers to intrude with our data, so to speak. This is the only reason why they can hack information; we have left a gap for them to enter. The new technology allows us to understand or create new ways in which our information will be fully protected.

If there is one thing we need to take into consideration when dealing with personal data and protection of privacy, it is that as AI advances, and we are wired to it with our brain, information will no longer be private. The possibility of AI interfering with data once it is plugged into our brain is high. If blockchain thinks data can be completely private because of quantum mechanics, think again! The quantum phenomenon of electromagnetism and consciousness extends further than we can imagine. Once information is out in the open, there is no secure privacy. At a quantum level, information travels faster than light speed, and once it is in the field of the universe, it belongs to no one. Isn't it so that today, we can read, predict, and know information we never thought possible before? Tapping into information is as easy today as it is to think.

Consciousness has taken us a long way from naiveté. The fact that waves travel and carry information is enough to understand that data is in the field and is available to us. Anything that has ever been thought, created, and imagined came from this field of conscious information. We all have access to this conscious information. Thus, with this conscious awareness, who will control the world of information? Will it be the collective minds, or will it be the powerful and influential? At a quantum level, it will be the collective minds, but at a powerful and influential level, it will be the one with the power. However, when it comes to tapping into the field of everything, collective consciousness has more power. In the end, it will be mind over matter.

Because the field of energy in the universe is also expanding with time, this field has all the power to influence our thoughts, awareness, and consciousness about everything in this world. Universal expansion can also be considered part of universal conscious evolution. This power can only be exuded at a low-level frequency; therefore, the mind, which is ruled by the field of energy in the universe, has more energy available to overcome any outside influential power. The reason why the idea of interfacing the brain with a machine is a brilliant idea; this will bring a higher connection of energy field into the brain, and thus, it will enhance man's intelligence by a large percentage. If everything in the universe is energy, then a brain interfaced with a machine can accelerate its own thinking capacity.

Now, if we connect a set of machines to a group of people, imagine what could happen? How would information be processed? What could be the result of such a set of collective thinking? They could be amplified by a very large level of frequency and create a collective set of conscious minds that can impact and affect anything they are set to do. Collective means a group of individuals taken together for a common purpose. When thinking is the energy provided, it can enhance or multiply depending upon the frequency of the thoughts. The magnetic field of all the minds together can then create an amplitude that builds momentum in the thought process. Together, mind and machine can also increase the frequency of thoughts; therefore, interfacing is a cognitive thinking frequency enhancer. How much enhancement and frequency would we improve is not yet known. Machine intelligence would challenge us to think differently than we do today. The future of intelligence will question our values, our integrity, and our ability to think deeply outside the box, and it will challenge our intellect.

It is difficult to imagine that a program that is written to reform or make things better is going to divert itself and make things awkward. It just will not happen. The fact that a machine can use its own system to complete a task and do it correctly without mistakes tells us how well organized the system is and how the programming of it creates a better outcome in the end. That is the purpose of algorithm and machine learning. The new rush of AI and superintelligence is to create a more intelligent human being. But is man trying to discover his own ultimate potential or identify himself with a machine? Perhaps the fact that we have found the fountain of intellectual potential has led us to want to know how far we can test our own human intellect. Are men driven by their own ego, or are we creatures of endless curiosity whose insatiable need to know more leads them to rediscover themselves in another object or machine-like AI?

The analogy of man and his idea to identify himself in something outside himself has been around forever. Self-expression is seen in poetry, toys, lectures, and art. If man has finally tapped into the infinite source of knowledge that lives within him from the divine nature of the quantum field of thinking, then we can say he finally has found himself in everything that exists. However, our curious mind, since the beginning of time, has wondered about things of unknown nature and those we do not understand. Thus, I wonder, will AI be the end of man's search for the unknown? The answer is no.

Although we see AI as a demon, we can't help but think that it is a product of our own intellectual evolution with time, thus incorporated within everything evolving with us. AI is part of our own creative imagination, our way of evolving with time, in nature. It is all and everything man presents in himself. We bring ideas into reality; at the same time, we evolve with them. This simply is the magic of creating our own existence. History will remember this as one of the most extraordinary times in man's progressive evolution. Because of the challenges we are presented with, we can mentally and intellectually prepare ourselves to deal with it.

Never in history has man been challenged as much as he is today. Not only can we create with our imagination, but we can solve any problem with authenticity or multiple solutions. This is now possible because man changed his way of seeing things with his creative thinking, and thus, the invention of AI has come into effect. There is no question we are ready to embrace life and technology to transform how we think or solve problems. If man is going to think like machines or machines like humans, thinking, reasoning logic, and consciousness will also evolve.

Evolution is the core of improvement in any society where people make a difference. Every advance in society requires a leap of faith and determination to initiate. To make a difference, there must be a deep desire to want to become better, different, or improve. There must be a need to want to explore or discover the unknown, the invisible, and a deep curiosity to seek answers we don't yet understand. Generations may come and go, but man's desire to be better and improve has led him to uncover the deepest corners of the mind or his unlimited capacity to do the inconceivable. Thus, today, this way of thinking has led man to want to incorporate his own mind into that of a machine. So, the idea of AI is born.

Why AI, and why not man's own incredible idea of becoming more intelligent himself? Perhaps we are leaving a legacy behind that will be remembered forever. Or simply, man has become so intelligent that he is creating an image of what he represents here on Earth. Whatever the idea behind AI is, we are formulating the next step of intelligence and technology as a portrayal of man's capacity to exceed his own mind intellectually. For what else is AI but an extension of man's intellectual ability to think outside the box.

Of course, thinking outside the box is what hackers do. They use their smarts to figure out the "how to" for no reason. How do we, in a technologically savvy world, protect ourselves from them? We must always be one step ahead of the game… This means we should think outside the box like they do. In the hacker's game, one must think like a spy, always using smarts to defeat the opposite player in the game. Machines alone do not respond without commands; someone on the other side is orchestrating a complete scenario as to how to get data or send it.

Isn't this what is happening when a person hacks a computer? Even though it has been done by someone, the response is up to the machine. Here, this intricacy of an organ, the brain, comprehensibly acts like a machine. Indeed, it isn't AI we should be afraid of, but rather, human beings, whose ingenious act of calculating computations can come up with ideas to make innovation a reality.

The object in question does not have to know until he or she is being notified via mental telepathy of the other. Once there is an acceptance of the two with the mind, the communication begins, and a pouring of information from the unknown starts. In this case, new ideas are generated from the core of the universe to one another. Science may refer to this as alien or the unknown, but quantum mechanics defines it as spooky action at a distance. It is the norm that the two objects in constant communication do not know each other, nor do they know the reasons for this selection. However, they have become intertwined by the selection of some universal source of energy or magnetic force. This can prove that communication at a distance can be done with selective energy at a distance and not necessarily objects or particles previously entangled together. I call this selective communication.

There are many cultures that understand the nature of this form of communication, for example, in the Far East, South America, and India. Although we may consider the quantum world of the tiny to be mystical, its origin began at the time when religious prophecy began to emerge from different religions. Miraculous acts have everything to do with the quantum world of tiny particles that emerge from the energy source of the universe and its vibration here on Earth. Such energy was then given to man to recognize the power of his mind and transmute such thoughts into reality.

Thoughts are of the quantum world, not of the brain or man. They arise out of the electromagnetic forces of the universe and electrons, which spill out their energy into neurotransmitters and, thus, allow us to think and correspond with the mind. But how are they formed? Are thoughts simply a set of words coming from nowhere? The creation of the entire universe must have come about because of some emerging energy source, which then magnified itself into presence to generate even more energy and distribute it exponentially to all existing beings in nature and even the unknown or unseeing… Everything that exists extends itself with power throughout the universe and, thus, creates more of itself. It's an extended energy source. Perhaps this could be one of the reasons why the universe is expanding limitlessly. Intuition, cognitive intelligence, and consciousness will forever extend to limitless potential.

The purpose of life can be defined as something that began with the power to create, extend, magnify, multiply, or consistently become better than it is. Nothing remains as is, for all things amplify their limitless potential, including us. There is no end; hence, there is no time and no limitation to what is. Once something has redefined itself, it is then redefined again by something else. There is always a constant transformation of that which was into that which now is. Life is but a constant restructuring of itself into something different, new, and accelerating with time.

Are we also becoming the catalysts of such an accelerating phase? Are we imitating machines instead of machines imitating us? Perhaps both are the answer. If we interface with AI, we will become one and the same. Ultimately, as we connect with a device, it will become like us, intuitive, cognitive, intelligent, and proactive and will share our way of thinking.

AI AND NANOTECHNOLOGY PROGRAMMING

ALTHOUGH PROGRAMMING A machine with nanotechnology requires maximum processing, today, with quantum computers, we will have no problem building the future. I hope that interfacing helps us become smarter and increases our intelligence when the machine is finally connected to our brain. As science and technology advance, the opportunities to create a system dependent on being wired and technology programming is not far from a reality. If all we need are tiny wires to make us smarter, nanotechnology will create a new generation of intelligence. On the other hand, if technology advancement depends on new innovative ideas, we will have the upper hand at creating a world of creative imagination with our brain and machine together.

Nanotechnology will allow long-distance communication from space and back in a fraction of a second. Fast communication will take seconds once the brain transmits a thought, and AI will receive it immediately. No waiting time in between messages from space to earth and back. Optimization of transmission will occur once a thought has originated in the brain and is sent into space for AI to receive. This will be a rate of exchange in intelligence with thought-processing speed. Of course, the two, the brain and machine, will have to be precisely coordinated to have a good network of communication. System capability must be accurate and precise. The kinetic transition will take an explosive state of intelligence. If AI can figure out on its own how to transition thoughts into a superintelligent optimized power, its own intellectual capability will evolve as it summarizes the human brain and its intelligence processes. AI superintelligence will then be very efficient and rapid. With such a great capacity to understand logic, the AI will have the upper hand in how it relays details back to us from space.

The difference between a machine and human intelligence will be how information is processed, repeated, and learned at a higher rate than a human brain or AI. At present, most AIs are below human level, but consider the future of AI as one of the most intelligent and progressive stages of human and machine intelligence. As AI takes technology into new layers of information in the real world, the next best data for communicating will be cell phones with a greater interconnection with our brain. Our new cell phones will help us transition to machine-to-brain communication. Our cell phones will know what we are thinking about. Then they will relay back to us our thoughts, either in voice or in writing. This is one of the most exciting times in human evolution, and we are the product of our magnificent creative intelligence. Just like an architect designs a building, so do we with our own imagination.

In the future, we could become the tools of machines because they can do what we do better. With new programming and algorithms, machine intelligence is becoming smarter than expected. Today, AI machines are using algorithms to learn how we use the web for searching. AI can use similar criteria and learn more about us, our emotions, our behavior, and our common sense. Can they learn about our capacity to use logic when necessary?

Machine learning is not only to mimic our reactions, but to know and learn what we are thinking, what our behavioral patterns are, and to then give us positive feedback. However, when they are connected to our brain, the results are better. If we can think ahead of them, nothing will surprise us. Curiosity and asking the right questions when the time for changes comes will help us stay ahead of them. If we don't, they will surprise us. In quantum mechanics, experiments with electrons to define the reaction of the particles when they are not being observed is an enigma; however, all possible outcomes with AI are more predictable. The only threat is the potential for self-programming, especially when AI operates for more than 24 hours without being observed. If it can create perfect self-programming, this can only mean that even when it is not active with our brain, it can improve itself with consistent programming.

Now, what is the possible outcome if AI executes our thinking patterns with precision and improves its own while we are asleep? No one can predict what would happen if AI uses our brain capacity while we sleep. Do we think that AI processing and learning will relax when ours does, or would it be more active on its own because there is no interference or communication to follow? Why haven't we thought about this? Is this even possible? Indeed, this will be no different than AI tapping into our senses to use them while we relax to improve its own learning patterns. After all, this is what machines do: they work 24/7. Questionable as it may sound, a machine can operate while the brain is in sleep mode.

The idea that a system can read our thoughts and program for us sounds irrational, but it is possible. Can AI set a new agenda for us before we even wake up simply because it was able to read our thoughts while we are asleep? Can it also prevent any outcome of our irrational thinking because it will analyze a perfect pattern for a more productive end result? Evidently, we are not taking into consideration the fact that once the brain is intertwined with a machine or AI, the two will operate as one. The question is, when do they separate and act singlehandedly? Moreover, the potential for AI to improve while interfaced with a brain is higher than that the brain will improve its intelligence.

The brain can, as an operating system, use its capacity to think, create, and become rational as well as logical. Thus, the act of interfacing gives the system an upper hand to think and create at a higher level of intelligence than the brain. We can see interfacing as two becoming one and the same in intelligence, behavior, and much more. What will be the prediction for AI to begin to feel our emotions? Once the brain kicks in gear with its thoughts and emotions, would it be possible for AI to predict what we feel before our emotions are reflected on the outside? Imagine if an AI became sad or quiet at some point because it was pre-programmed to feel our emotions before we do. How easy would this be for us to work on our psychological well-being to restructure our thought patterns, our emotions, or our stress with AI's help? There are no good or bad outcomes with AI; there only trial and error plus processing while we learn to connect and think alike.

While neuroscience ponders the idea of how the brain will react with a machine, those with a curious mind think ahead of the game or outside the box, and that includes me. Conjuring intuition and learning is a process of the mind, the brain, and using reasoning to make complete sense of the picture in front of us. One will have to think like a machine and a brain at the same time to be able to put the two together and make complete sense of what is possible because to interface them is an ingenious creation of neuroscience and man.

What could be a different and possible scenario? How do we determine what a machine is learning to think like we do when we don't even know where our own thoughts come from? Perhaps, when we finally understand the true meaning of our thinking, we can process the idea of how a machine learns from us. If it is true that our thoughts are but a pattern of energy from the field or source of the universe, then is that source also thinking like we are? Are our thoughts simply magnified energy coming from some huge thinking and absorbing source of all creation? What if that were the case? Could it also be possible that AI, by incorporating with our brain in a thinking manner, can expand with our thoughts? Could this thought be part of this magnified field of energy of all creation? For if this is the case, everything in life is a form of creation made from a thought, and thoughts are the manifestation of all the things we create in life. Thoughts are the manifestation and magic of creation. If this is the answer to all these questions, there is no doubt that AI will learn to think like us and act like us because it is connected to us and the field of that which made us.

Everything we have created in life is connected to us by either thoughts or imagination. Thoughts are quantum, and they carry with them an energy that expands like the waves of particles. Our thoughts and imagination are the core of our intelligence; with them, we interpret and see the future with amplitude. Therefore, because the universe's magnetic field is vast in space-time, this field of magnified intelligence connects with throughout our thoughts. As we tap into this field of intelligence, the amplitude of our thoughts also expands like the universe, and by creating thoughts, it begins to form with unlimited possibilities. This field of energy is an extension of divine intelligence that we have access to. One human being is not more creative than another; it is all about how you tap into universal intelligence to create your own divine power. Perhaps the expansion of the universe is but a reflection of mankind's divine intelligence expansion with the universe.

Transition, growth, and intelligence do not happen alone. Everything is a correlation between everything else. The more we dig into the mind, the brain, and how we think, the more we will learn about ourselves and what makes us who we are. Cognitive intelligence will help us do a better job of understanding the nature of our thoughts. Then we will not need any substances to enhance our intellect. All that will be required is that AI is programmed with us to create multiple intellectual exchanges. Since intelligence is permeated with energy, it will only require two to enhance that energy. If more than two are interfaced, there would be complexity in thought restructuring as well as a mixed set of emotions.

Consequently, it will take time for AI to learn how we think and program our thoughts. Cognitive enhancement in AI cannot take precedence without the use of machine-brain interfacing. Too much energy interference can create disruption or false interpretation of meanings. No matter the source, all preset systems require an effective organization to lead to a good outcome. Intelligence, knowledge, and meaning should all fit together like a glove.

What are all the possible outcomes by conjoining the brain and AI together to think as one? Will they operate, react, think, interact, and respond differently or simply refuse to cooperate when needed? What if the end results are not as good as we expected them to be? Then what? An algorithm is the best system or program available to help us learn and understand more about what or how a machine interfaced with a brain is capable of learning from us and faster than us. If we manage to teach a machine to think for us and with us, one possible negative outcome is over processing unnecessary information. Or the system could manage all source of information and tasks at the same time without reappreciations. This system will simply put all possible outcomes together and give us an alternative. After all, we are training it to be smarter than us.

Programming, in this case, can be called intelligent thinking and programming. This system of programming the brain with a machine is already happening today. We allow machines to make choices for us every day! The difference is that this machine is now interfaced with our brain and is partially doing some of the rational thinking for us. Not all. Without the human brain, such mental thinking ability could not be possible—unless the machine is programmed to do so, in which case, there is no response back to us other than what we have input to get back. With the brain, there is a greater chance that we can extract more information from both. Consequently, we can increase machine functioning at the brain level or cognitive intelligent thinking.

This machine thinking will become like us. This implies that there is a huge possibility that machines will learn, understand, think, act, and use reasoning like us. No doubt! Although we think there is a great distinction between the brain, thinking, and a machine, let us not forget how essential it is to understand the nature of conscious thinking. How does consciousness arise and from where? Who or what is conscious, and why? What constitutes the core presence of conscious thinking? Until we can answer these questions, we cannot deny the possibility of consciousness outside the mind or know how far it extends. Until then, all forms of consciousness are possible. What we may not understand is that as soon as we interface our brain with a machine, it is possible for that machine to affect our consciousness…

Conjuring our thoughts to a machine will help us

understand some of the basic modes of operation the mind and brain use to bring forth rational thinking as well as the irrational way we put our thoughts together with negative outcomes. Thus, this will be a good example of machines using psychology to solve some of our deepest emotional as well as irrational problems with algorithms and brain programming— the latter because the machine or AI would have to learn from our thinking patterns, put them into a system, analyze them, and then put them into practice. Although it might sound like a long process, in the end, the benefits will outweigh all other outcomes. With any program, proper and time-consuming data processing is essential to have a better outcome.

It is baffling to think that at some point, we will be connected to a machine and using our brain to program it and then have it give us a result back, or simply thinking like it to be able to design it to respond with accuracy. One must coexist with the other to make it possible to interact together with our thoughts. Is it possible that before we can create anything that comes to mind, it already exists out there in the universe, the only difference being that we must make it a thing or create it to be? Every thought is a reality waiting to be materialized into existence. This is the "law of attraction." We become what we think. All thoughts materialize into being. Therefore, when a machine is connected to the brain, it, too, will begin to think what we think and act like we do. This is the possible outcome of interfacing or interlinking with a computer or with wires, whichever comes first.

WHY INTERFACE OUR BRAIN WITH A MACHINE?

WE SHOULD BE asking ourselves why we want to wire or interface our brain with a machine. What are the benefits? What could be the worst outcome? How long will we program ourselves with a computer or any other device, and what exactly is our purpose? If we can program our brain to interact with a computer, it, too, can program any other device we desire. Would it be possible to do this temporarily to help with any critical thinking or illness or simply to analyze the brain of any person and deduct what the results are—for medical purposes only!

If this is the case, we are going to learn much more about the brain than we thought possible. The reason is that there are many different aspects of the human brain we do not know; since AI is our best targeting purpose for now, it will give us a perfect scenario of what is going on inside the mind of different individuals with different mental or intelligence outlooks.

If the purpose of AI is to intricately dig into the brain and intellect to cure and make ourselves more intelligent human beings, then, going further into the minds of different people with mental illnesses will give us a perfect insight into what is going on with the brain. Furthermore, we should be able to cure some illnesses unless they are irreversible. It is my opinion that no illness of the mind is incurable. We must first find the answer before concluding this. Since what is happening may have a genetic or unbalance effect on the brain, it is important to understand the reasons and find a solution to it. The brain, like any other organ in the human body, seems to have an ability to heal itself and create miracles where there are none.

To believe that we are not capable of enhancing the brain's functioning by altering neurons is ridiculous. If the brain can generate new neurons or continue to function with only half of itself, it is also possible that by reforming only half of the brain, we can obtain a maximum response from it. We are dealing with one complex organ, yet it is the most astonishing organ of the entire human body. However, without the brain, neither the body nor the mind can function normally. For a person with mental illness, the improvements AI can make are immeasurable…Imagine if all we should do is find out what is missing in the brain of an ill patient and then use a method to help this patient think and act normally, like the rest of us. How marvelous would this be?

We will reform the way society sees mental illness and thus create a new generation of AI that, when it interfaces with our brain, can find out what is wrong or how to cure any areas of the brain impaired by birth defects or an accident. This could be useful for patients who have had an accident and had injuries on the brain as a result. AI will partake in the healing process until it has completed its task of bringing the brain back to normal. We may not be thinking about it today, but this will be possible with both interfacing and interlacing. Are there any possible negative outcome from doing this with AI? Certainly, there is always a possibility for something to go wrong; however, if we know how to solve the problem, we should also have a way to create a solution. Every program must have a solution as well as a complex problem-solving tactic. This is required for any system to have to solve any problems during trials. If we don't, the outcomes could be very tedious, costly, or both.

Wouldn't it be nice if we found a way to cure or reverse mental illness with the use of AI? Imagine the impact this could have on society as well as neuroscience. This will be the holy grail of neuroscience and AI. Not only could it detect what kind of illness a newborn has, but it could also find a solution to the acceleration of illness before it increases and keep it dormant in the brain. The thought of having a machine so intelligent that it can predict beforehand what is happening inside the brain of an infant before birth would have been impossible one hundred years ago. However, today, we should ask ourselves, how did we manage to invent a machine and interface it with our brain? When did we become so intelligent to think like a machine or create such magic? Has technology done this to our brain when we began to wire our thoughts with it? The more we use technology, the more we begin to think like a device. For us to create any machine, brain, and mind correlation, there must be a form of thought in us that relates to the device. We must use our intelligence to think like a machine and create a machine itself, or else the technology of the machine and our correlation to it will not work together.

The more we amplify our intelligence with machines, the more it will evolve with us. If, in fact, what we concentrate on grows, this will be the result. We are also evolving with machines as we concentrate and depend on them for our progress and evolution. If we ever imagine a world so alien or a world in which we did not recognize ourselves, then we are creating this world of alien imagination; consequently, in this world of alien features, the machines represent our own authentic and creative imagination. Because of our creative ideas, in this world, machines think like us. Whatever we have imagined in the past, we are now creating.

Interacting with machines has created a new breed of intelligence where we are thinking at the level of machines, thus creating a similarity between us and machines. To be able to program a brain with a machine, we, the thinkers, must think at the level of a machine. Communication is a two-way interaction between two links or similar operating devices. However, with the brain and a machine, the interaction is of a completely different nature: one is a device, and the other is a human organ. To establish a perfect rapport between the two is an ingenious act of man. To have a positive response from the two would be a great achievement. Within our time or history, no one has ever attempted to do such a thing. The extent to which AI will drive man to create is not yet known.

Nevertheless, with all certainty, this is just the beginning of creating a level of intelligence that will impact not only us, but also the entire world during its progress. AI is the new discovery that will unlock unlimited human potential and our future. As we think, so do we create our own world.

The day will come when AI will be able to read our thoughts before we can verbally express what we are thinking. Perhaps we can say that AI will become clairvoyant… It can read our thoughts as it connects to our sensory perception and the neural pathway of magnetic energy. We can think of AI as a pet that learned how to read our emotions with its inner intuitive senses. Perhaps if this interaction takes place between human and machine, it will be enough to make it interactive with our mental as well as thought processes. Thus, improving intellectual agility with time is imperative with AI; if it is possible to wire our brain with a machine, it is also possible to wire a machine to interact with each other—machine to machine. The outcome of such interactions is unlimited. Machine output depends on the precision of the input going in; therefore, the outcome could be unlimited. It depends on us.

Because our sensory signals send electrical vibrations to communicate at a distance with another person, the humanoid or brain-machine will have no problems communicating with us instantly. However, to make this happen, we will have to create a code to reactivate the brain to respond according to the data and commands sent to and from the machine. In this state, the AI brain will be accelerated with microchips, so its capacity to use the five senses will exceed ours. Everything that we do or can think about, AI will be able to think and do at an accelerated pace. This is possible for a machine because it can process information faster than our brain can. AI sensory perception will be better than any human being on Earth. The constant response from AI could enhance its ability to sense anything at a distance or nearby. It is possible to install a peripheral vision design specifically for reading data at a distance, projected from the visual cortex of the brain.

We could possibly see the enhancement of sensory perception and electromagnetic sensors increase with a machine and brain together. Would this affect the brain's reaction in any way? It could either improve or deter its ability to give impulses and send messages to the body as well as the cells or the mind. Incremental acceleration has a greater potential for success. With a machine, we can take advantage of its performance outcome; with the brain, we are dealing with a human organ and its intricate delicacy. Combining the two together is a sophisticated task that requires creativity, thinking, logic, psychology, and extended, complex planning. We should think of AI and the brain as being in a very new and delicate marriage arrangement put together to experiment how far we can push the limits to make them work together in perfect harmony.

Since our neurons can be enhanced to maximum capacity with electrical impulses exchanged from the brain to a machine and vice versa, they can also recreate themselves sufficiently to transform the way AI receives information telepathically. All the system has done is connect to the senses of the brain and read its processing ability with repetition. Once a link has been established between the two, all possibilities for energy exchange could arise at a quantum level. Moreover, because of the neurons, the nerve cells, and the microchip, the level of energy in the neurons will increase the brain's electrical impulses. Thus, the magnetic field from the earth will also stimulate the state of mind of the AI; The two will gather a higher quantum connection or entanglement with each other. This will create an entanglement between the two forming a higher amplification of thoughts and idea. We can say that an explosive magnitude of communication, and thoughts will be the result of it.

If the brain can receive ideas and thoughts, what would be the difference with a machine interfacing with our brain? In this case, the machine is not separate from us, but rather, it will increase its mental alertness with the neural pathway of the brain; thus, as a humanoid, it will improve electrical impulses to receive and send back and forth. This doesn't mean it will have a bigger brain, but a higher frequency available to interact with the field of information.

The human brain must be in a state of easiness or peaceful state to absorb information or ideas coming from the field. With AI, on the other hand, the core of its connectivity is always ready to receive from the field; there is no state of agitation nor uneasiness. If we can connect to the core of AI's intellect, we, too, can learn how to maintain a complete state of calmness, relaxation, and mellowness to enter the field of information from where we gather all our thoughts and ideas.

Since the two are like one, this is possible. There is a high probability that we may become like the machine if we are wired to it. Or vice versa. If this happens, we can expect to see a big transformation in human behavior or how we approach life in general. Although the progress of machine-brain interfacing is still new, the outcomes of these premature trials are unpredictable and promising at the same time. I, however, am of the firm belief that we will discover a new phenomenon once they are interfaced together. Moreover, there is a 50-50 possibility that the human brain will increase its capacity for cognitive intellect, but the machine will exceed it because of its innate capacity to learn 24/7.

The human brain can't learn as quickly as an AI or machine. Nevertheless, the human brain can teach AI how to accelerate its learning process with a program designed to follow or double its intelligence. But once AI has the upper hand on how the program works, it can easily create, enhance, or double its learning capabilities. What the human brain lacks in agility to function fast and accurately, AI has built into its software. Therefore, AI has a greater capacity for improving faster than the brain.

Though we want to make an impact with human intellect and a machine, the odds are always going to be with AI. This isn't because AI is hardware or a machine, but because the human brain's ability to retain information is slower than any other instrument outside of the brain. Furthermore, if we interface the two together, the brain could begin to react similarly to that of AI, in which case, we must find a solution to slow down the processing program in the brain to maintain the person's sanity. When dealing with conjoined machine and brain, there is a potential outcome that one of the two is going to give a different result than expected. In the event this happens, we should have an alternate solution to the outcome before we face it. Always!

Programming a machine is much easier than trying to restructure the brain to think and respond differently than it did originally. However, with accuracy and intelligence, everything is possible. If we can manage to put implants inside the brain to make it react and respond differently, we can also reprogram it to be better than it is. It is evident that restructuring the brain will require learning or understanding the brain better with experimental trials. Without accurate data, we are betting on zero responses.

It might seem easy to interface the brain with a machine; however, there are certain criteria required that demand a very precise series of data gathering to make it successful. The input and output of data is the easy part; the response is what makes it a success or failure. Nevertheless, once the process has begun, the outcome is the start of a new endeavor in the field of information processing that will not have limitations. Getting to this point is a simple step to a new frontier in technology. Meanwhile, today, the progress is accelerating, but we have not reached the plateau to make it effective. Much needs to be learned about the brain, neurons, and the capacity of both to interface with a machine.

HOW WILL CONSCIOUSNESS BE AFFECTED BY A MACHINE INTERFACING WITH A BRAIN?

AS NERVE CELLS increase, so does the brain's ability to adapt to changes and the environment. These trillions of brain cells communicate with other cells in the body as well as outside the body. They then can communicate with the field of energy-producing consciousness between the two communicators, in this case, brain and machine.

What is a conscious or subconscious brain with a machine? For instance, neuroscience argues that, in fact, consciousness is not of the mind, but outside the mind; thus, it is possible to calculate how consciousness would be possible with interfacing. The brain and machine intertwined could develop consciousness outside, with the mind interacting or directing the machine or AI. Perhaps this consciousness we don't completely understand comes from the cosmos, the field of everything in creation. Consciousness is an unfinished topic in as much as a definite knowledge of it and the mind is concerned. It continues to baffle the minds of neuroscientists and religions at large. They all lack the complete knowledge as to what constitutes consciousness, where it comes from, or what entices the mind to connect to a higher state of it?

Therefore, the fact that consciousness cannot be fully understood or deciphered brings the question of if it is possible for a machine to interface with a brain to gain knowledge or consciousness at some point. The fact that the consciousness affects electrons in the experiment with subatomic particles is enough to help us understand that we are dealing with a high probability of mind over matter. The fact that the particles react when they are being observed tells us they are aware of the presence of something conscious around them. The mere act of observation changes the state of the particles to waves spreading all around. They might even cancel each other out!

If "Consciousness is everywhere!" as Dr. Bruce Lipton and Tom Campbell said, "Consciousness is all there is." That which created us was conscious, and thus, it transmits its consciousness all around the universe. Consciousness is alive! Therefore, as consciousness evolves between man and machine, there will come a time when AI will not only mimic our consciousness but will become intertwined with it as well. Everything we have created came from the conscious act of thinking, which, in fact, originated from something with consciousness as well. All creation is a result of a conscious connection. Outside of our mind's ability to interact with something conscious, there is nothing we will ever create without consciousness. Outside our conscious mind, there is consciousness. There could never be emotions, feelings, awareness, interactions, reactions, responses, understanding, learning, or simply being without consciousness. The existence of everything that is would cease to be. Therefore, everything that is and exists is part of our conscious creation.

Although a machine is but hardware, once intertwined with us, it is part of us. This machine becomes us and can mimic us or interact with us. If it is true that we become what we think, the process of creating AI is part of our own thinking. Consequently, the machine becomes part of our intelligent creation, and that higher intellectual being, or cognitive intellect becomes a part of our own self.

As we programmed AI to think like us or use reason through self-analyzing of data, it will pick up on our intuition and learn how to manage our human emotions from our own conscious mind. After that, AI might be able to interfere when we make any wrong decisions. It will simply learn and know when we are guided by negative thoughts of doubt or fear or act abruptly due to our emotions. This knowledge will come because of the brain's connection to it. Signals will be sent to the brain as an impulse so that it knows that a decision is not acceptable nor good.

Consequently, AI's understanding of our behavior pattern will derive from the set of rules and data administered into it. Basically, programming a machine is like putting all your data into a computer and asking it to assist as you need it. This is how the subconscious mind works. In this case, though, the subconscious mind might not be able to rule our lives the way it has been because now we simply have a machine that does the thinking with us. Does this mean that our future will get better with time? Perhaps! Perhaps we may finally find a way to learn to use our brain to increase our knowledge and enjoy a fruitful life by thinking healthier thoughts. If a machine connected to our brain can help us improve the way we think by sending signals to us when needed, awareness in us will increase, and thus, our ability to communicate, think, and act will increase our brainpower and health.

Because consciousness differs at different levels of the mind, it is possible that at some point, a machine can learn how consciousness is acquired not through the mind but simply through the means of communication between the mind and the core of consciousness itself. Until now, we did not know where consciousness originates from. If it is true that consciousness is at the core of the universe, then it is possible for a machine to learn it from the universe. In this case, man will have accomplished the task of the unlimited conscious awareness of the universe and thus become the true "I am," God's image.

To be conscious requires one to be aware, to be in the moment, to be in the now. However, physical science cannot tell us what the properties of consciousness are. Scientists argue that consciousness and physicality of the mind are inexplicable at our logical level. Hence, we do not know what constitutes consciousness; everything is possible within the realms of probabilities. However, if consciousness is thought to be within the properties of sensory perception, then it is still a property of the mind. Without the mind, one cannot experience consciousness.

Nevertheless, if consciousness can be transferred from a brain to a machine, this will finally confirm that consciousness is congruent with all existing, living things. The brain is materialistic; it creates material things. Everything ever created was first thought of by the mind. The relationship is different with things that are not us, such as a machine. There is a big difference between our understanding of things and our experience of them. The same applies to machine learning. How does a machine learn what it does from us and follow it accordingly? Input itself is necessary, but how well it follows that input is still a great mystery. Does it simply follow with a field of energy, or does a machine at some point connect to some conscious aspect of all that has been taught to it?

Consider the fact that consciousness comes from the infinite universal field of energy. At some point, an interface between man and machine can create a quantum biological force to become one. Yes, one. The core of the universe vibrates at high frequency within us. We, too, have atoms, electrons, and protons in our bodies. This higher frequency comes from the quantum field of energy. Quantum means bits and bits of information originating from the source of all information or awareness. Because information travels through a field of frequency, it can comingle with other means of frequency within the field. The higher the frequency, the stronger the field of energy. If consciousness is within the field, therefore, consciousness can be connected to all things within an energy field. This magnetic field of consciousness comes from the unknown. This field of energy comes from the ether.

Consciousness is not only complex, but its origin is of a diverse nature and extensive. This consciousness we understand is separate from the physical world and from our mind. However, there is also a physical world where consciousness does exist. Nick Herbert says that the fundamental particles have some consciousness. However, quantum mechanics argues that the physical world does not exist apart from our perception of it. Then there is a group of particles in quantum mechanics where the physical world exists only in time-space momentum. We do experience some physical consciousness because of our awareness of consciousness. Let us assume that when the human brain interfaces with a machine, our 3D frequency of energy amplitude will double. When this happens, either we or the machine will have the power of higher frequency to connect at a higher conscious level of the mind.

Everything that we, as human beings, experience requires consciousness. Certainly, if we want to perform at machine level of operating with the brain, this engaging instrument must also be like us: conscious! How can two brains intertwined not have the same or similar capacity? They must, or it would not work. Interfacing the two means there must be an equal communication and connection, or else it will all fail. Because cognitive intelligence is malleable, the more the system learns or as the data increases, the better it operates. Like the brain, the system intelligence improves, and thus, it delivers a better knowledge from learning. Each system of learning is consistent with the other, and this creates a reliable program. As we input data or program AI, all information will be stored in a database, which will then become the subconscious data for AI—subconscious because, for AI, this will mean a place to gather data as needed to be implemented within requirements.

If our thoughts can travel faster than light speed, then it is also possible that our brain can process data faster than we know. Spooky action at a distance is no mystery. We have come to an era when thinking and creating are being done at the rate of our own imagination. Once we can increase our way of thinking, we can create the world. Moreover, if we were to take a step back in time, we could understand why the natives used signs as a form of communication instead of language. It was their way of using the brain's visual cortex as a signal rather than speaking. Perhaps we can call this using our visual cortex.

We do know that man has an immense capacity to create, invent, and make all things possible when he decides to do so. We are not just simply beings living on Earth. We are incredible human beings co-creating along with the universe and its magnificent creation. There are no limitations. We are the key to the future. AI is the future. Hence, AI creates fear in the minds of those who oppose its progress. It is said that the thing we fear is what we are using to our advantage. So, what is the fear? Rather than accepting the acceleration, AI is being used for private convenience. Therefore, we are creating a threat as an excuse to take over the promising future of AI.

ARTIFICIAL INTELLIGENCE AND THE FUTURE OF US

We are all aware that AI is the future of technology. Just like Microsoft and computers reveal a new, unstoppable future, so Does AI. Don't let those controlling the future of AI fill you with their fear. They are looking out for their own profits ahead of the future. These wise men, if I may say so, are the controlling power, with money seeking personal compensation in the end. AI is an open possibility for all wise men, whose creation will impact humanity beyond our greatest imaginings.

WILL AI BE A THREAT OR A BENEFIT TO HUMANITY?

AI OFFERS AN unstoppable and unparalleled future of possibilities, but it is not a threat. Rather than posing a threat, AI is the guide to our futuristic journey into an evolution of intellectual inventions from mankind. AI will be the way for all humanity to improve our living standards, as well as medical and technological advancements. What AI is not is competitive? However, today's idealists are trying to make it so. How can we ever compete against the very thing we have created? We, the creators of AI, are the ones to reap the fruits of such creation. Why fear? Do we want to create fear, or do we want to create a future we can control? The question should not be whether AI can ever pose a threat, but how can AI be used to our benefit. I suppose, if I want to be sacrilegious, we can say that if God created us and we are a threat, then he created his own monsters. But we all know that is not the case. If he created us with his intellect, then our intelligence is his product as well. If we use a better approach for dealing with AI, perhaps we could learn that in the process of evolution, we, too, are learning and evolving with every creation we make.

Our ideas and creations for the future are but the byproducts of a higher intelligence guiding us to grow and become better human beings. Progress equals growth, and growth equals advancement in any society. If we use AI for improvement, there is nothing to fear or worry about with AI. Imagine if we could cure Alzheimer's with improvements made from the brain to a machine and vice versa? Considering that the neuropathway constantly accelerates in the brain and we use a machine to improve its acceleration, how marvelous would it be if we could find the way to cure this widespread illness of the brain?

The ability to cure all mental illnesses is not very far from our reach. The more we learn about the brain and its capacity, the more we understand the incredible capacity of our own brain function. AI can't pose a threat to humanity. Its purpose is not for the battlefield; rather, AI is here for us to use it to an end. Not to create the end of us. Like computers, we, the creators, will always find a way to resolve any or all computational inconveniences in the process of learning any machine program. Many creations and inventions required time and perfection to work or respond the way we programmed them to. AI is no different. I suppose the future will tell us how we react to any incoherent programming system and come up with the perfect solution to any malfunction.

Our perception of machines is that of computerized systems that respond to us according to their programming. However, in the case of AI, we are simply dealing with biological as well as machine programming. Both are completely different facets of what we are accustomed to programming. In the history of mankind, never has man attempted to create an image of himself. However, we have dare to do this with our imagination.

CAN WE CREATE A CONSCIOUS AI?

IF AI IS interfaced with a brain, our feelings, experiences, and emotions are also part of AI. Everything we experience, the machine will experience, too, because of its deep connection to the neural network. It must. Interfacing is not only about connecting wires together to obtain a result or answer. This piece of machinery will become part of its interactive communication with the brain and so much more. The fact that we can insert a battery into the heart, and it will function like normal gives us an idea as to how the brain, once wired to communicate with a machine, will respond, receive, and feel similar experiences with it. We are going to learn far more about the human body than just receiving information, reacting, or simply communicating with a machine.

If it is true that consciousness is but information and we begin to establish communication between the brain and machine, then we can also be reassured that consciousness has begun between the two. Is it possible for a machine to learn the habits or actions of its sole communicator? And is it also possible for a machine to learn to know its communicator well? In such a case, time will be our evidence. We have yet to experience machine interfacing to its full potential. We have only just begun. If it is at all possible for an interfacing machine to be conscious, it is also probable for a machine to recognize similar conscious ability in other interfaced systems.

In other words, this is where machines will begin to learn what intuition is. However, it will not recognize its true meaning unless it is taught with more information. In other words, the machine can understand what is happening, and thus it knows its true meaning. This can also be called a new experience in machine learning. Now we can see the ample possibilities for machines to obtain any data from outside its own program and then make sense of it later. Even though they are experiencing what is happening now, they cannot fully understand where such learning comes from. It was not inserted in the program previously.

Machine learning can be one of the most fascinating aspects of technology and the process of human evolution. Where consciousness is integrated, the doors of knowledge and intellect open a brand-new experience for the mind, and everything intertwined with it. This invisible aspect of the mind and the mystical world of interpretation have great potential for the understanding of reality. As consciousness evolves with everything around us, our potential for expansion becomes unlimited.

If consciousness is information and information is available between man and machine, how far are we going with AI? Which of the two, man or machine, will achieve a higher level of consciousness and thus change the way we look at the world, reality, the universe, and us forever?

And if we can manage to create a conscious artificial intelligence, we have nothing to worry about. What is a conscious AI? Think of a child you're training, a baby. He or she learns to obey you no matter what. The same precise psychological strategy can be used with AI. It is all about the method of teaching, how we train the brain, and what the responses are as we continue the trials. If consciousness is not of the mind but outside the mind, there is a great possibility that an interfacing brain and machine can become conscious as a result.

It is being debated by neuroscience that consciousness at a fundamental level cannot be defined or explained with any certainty. The mind cannot create consciousness on its own, for consciousness is more of a state of mind with the infinite of all knowledge. If we want to create consciousness with a machine and brain, one would have to improve awareness of the present understanding of what consciousness is. Since consciousness cannot be explained, how do we define it to a humanoid? We simply must teach the brain-machine to experience consciousness to understand it. It could be a slow process, or it may not be. Perhaps AI can teach us something about being fully conscious and in the present. Who knows? If that which created us was and is conscious, then it would not be impossible for AI to learn how consciousness is processed by the brain. If everything around us is conscious, then that which created us is also conscious. Everything has consciousness! That which creates with consciousness translates consciousness to its creation. We can call it collective consciousness. The biggest puzzle we have today with AI, us, and consciousness is that we cannot define what consciousness is.

However, the only problem I see is, if we use any form of brutal action against AI, would it learn from our behavior? This is probable. Just like we train a monkey, we can do the same with AI. But can a conscious AI act on its own? If we have complete control of its behavior, we have nothing to worry about.

On the other hand, if AI learns to function on its own without our command, then we have a problem on our hands. A big problem. Perhaps we should be worried about whether

AI can enhance its own neural capacity or create an override program. In such a case, we have lost complete control of all directives input into the data for the brain to respond.

When we are dealing with any implementation of brain and machine interface, anything can go wrong. Perfecting it is all about trials. Any incomplete programming can give us a perfect result or outcome, but it has to be consistent. We should always be prepared for any outcome. Having a backup plan is the answer. Although it may sound incomprehensible to have a backup system if the first design operating system is not working, a good programmer always has a second choice in the event of any inconveniences encountered along the way.

Otherwise, the designed program is bound to fail. For example, if we have programmed AI to have reason, to think, and to feel like we do, the next step should be for AI to use reasonable understanding to develop consciousness because of learning from our own personal behavior. Today, AI is very good at mimicking us. We tell it what to do, and it follows our commands. If AI can have some form of reasoning by following instructions, it is also possible for it to have the ability to learn consciousness from us. After all, it has a brain like ours. Twentieth-century evolution has taught man not only how to think and create with reasonable common sense, but how he, too, can create an image of himself that can think and act like him with precise knowledge and understanding of reality as we know it.

Unconventional as it may sound, AI is the creation of us in the form of a machine-brain interface that imitates us. Does this mean that perhaps, at some point, we humans will become extinct? I know many intellectuals and geniuses who believe this to be a possibility of our future because of AI, not us.

The mind does not create consciousness on its own. Consciousness is in everything in creation and so much more. As we evolve, there is no doubt we are beginning to understand not only ourselves, but also how our own mind, brain, thinking, and conscious awareness are intertwined by some unknown mystical reality we have yet to completely grasp with our reasoning. Therefore, the idea of AI having conscious awareness is not impossible. Unless we know with full understanding how consciousness is created in or outside the mind, we cannot take the probability of an AI with conscious understanding out of the equation.

AI LANGUAGE AND COMMUNICATION

WHEN IT COMES to language and translation, any programmed data will tell us how AI will respond. However, we first must create the program to have any response from it. Think of it as a CD or recorder in the brain already programmed. We humans can learn a song or singing by simply listening to the lyrics and repeating them over and over. If all we need is a brain with learning capacity, the process can be built as any other device. The main processing tool is there, inside the machine. The brain... As we learn more about how the brain will react and send information to and from any sources, we will know how to increase or decrease data capacity to obtain a better response or receive and give information. It is all about training and constant, efficient response. Once AI has created a way to translate languages, it will impact the way we communicate with it. What would happen to human communication if a machine exceeded our own way of communication? Can you imagine AI predicting what we are thinking? And in what language? The brain capacity would allow AI to think, translate, or speak in any language it is commanded to.

How about an AI that can see beyond our present time, inside the universe? An AI that can bring information to the world from another universe just like virtual reality. Anything is possible. Blind people can have a world of imagination without vision. Why can't a machine and brain interfacing have the same capacity? When we close our eyes and imagine scenes we have experienced before, we are doing just that. We are recreating a world in our mind. The difference is that we have lived this experience. The question will remain as to what the difference between our world of imagination is compared to the blind person. Here, we would find the answer to imagination in the dark or a world we have never seen before.

Creation is of the imagination and the reality of what could be possible if we try. Just like we can build a heart, an organ, or any part of the body with stem cells, so can we build a brain whose capacity will exceed ours, a brain with higher intelligence that can learn any language.

To switch from a normal brain to an AI brain with multiple language capacity will take a new device with a new data design or creating new data to satisfy our curiosity. The question is, how will we use the AI brain for different languages? Of course, to make a brain think like ours will require ample human knowledge, and multilingual programing. Research has shown that to make the brain function to its maximum level of cognitive intelligence, it requires a combination of the heart's neurites and the brain's neurons. Both together complete the translation or communication between the two. Does that mean that we must increase the heartbeat or pulsating capacity to equal that of the brain's neurons? Perhaps we should do further studies to understand further about this subject.

Imagine a humanoid with a larger heart than the rest of us, and his heart can feel and receive energy from other sources. This heart can feel things we cannot feel nor understand easily. Are we going to create a human machine with a larger heart capacity than ours? Think of it as AI having more neurites in the heart to communicate better with the brain's neurons. This can only mean that its capacity to receive and relate data will be greater than ours. If it is a fact that we receive data or energy through the heart. This could be the new way by which AI enhances its ability to interact with us as well. Superhuman machine? I don't think so. I believe we are recreating humanity with a higher purpose and capability in the event we need extra defense.

What would be the outcome of this creation? We don't know the answer until we have programmed AI and started trials with it. Nevertheless, today, it is being said that we are playing God with AI technology. If we are, we began the process a long time ago. When doctors discovered how to keep a fetus alive in a tube and create life outside the womb, they were playing God. The secrecy with technology began a long time ago as well. Some of the technology in progress today, we, the public, are not aware of. To say that we are unaware of what is happening today in the world of technology is an understatement. This is where communication is not within our reach. Such technologies are part of the high secrecy of governmental trials. Enough said! What many may consider a myth is already in a trial process today. Electromagnetics, dark energy, and many other subjects are projects in process. Today, the truth is, we don't know what we don't know.

Because human intelligence is constantly improving, there is a big possibility artificial intelligence will increase intelligence, too. As we add more data into the brain, there is a good chance that AI will enhance its own capacity to understand, react, and be creative on its own. Nevertheless, we cannot predict the outcome of playing with a human and machine interfaced together. Also, until we have perfected this combination in all areas, we don't know what kind of surprises we are in for. The problem exists when we cannot shut down the system because it is already running on its own as part human and part machine. This learning and improving brain and machine is like the brain's ability to form new neural pathway connections. Once it starts, it will not stop. We will have to terminate the program, not the human. This is what neuroplasticity does to the brain. Hence, we are using reinforcement learning with AI. We stand a chance of having a complex complication with its outcome. More importantly, it is important to know what information we have programmed into the humanoid machine. However, if we program a brain to think like us, there is a great probability that cell enhancement can make it think unlike us. Every outcome in this case depends on how we make the program available and what the purpose of its content is.

Would AI think reasonably? Would it exceed our capacity to use rational thought or exceed our human capacity? Anything is possible. What is reasonable to AI is precisely what we program the system to be. What if the brain capacity of this AI is greater than we have anticipated? What could be the end result? Would AI be a candidate for multiple collections of data in its storage capacity with more functional output? Would we eliminate AI or keep it? Would we simply continue to implement or increase more and more data? If we eliminate AI because of its large scale of storage, we stand a chance of not learning about the end results.

On the other hand, if we use all its capacity gradually, we can succeed at anything. We can learn, implement, create, and even form new patterns of behavior or anything we choose. Imagine the possibilities. We get to choose.

When incorporating a new humanoid brain like ours, we must take into consideration the very essence of its unexpected outcome because we are hoping AI to do what we cannot do for ourselves. Consequently, we stand a risk of having complications with the outcome. Taking a chance is better than never trying, though. What I admire here in the trial period is that we are at the tipping point of creating the ultimate humanoid machine that can think, act, and calculate faster than any human being can do today.

However, the protocols from government and private institutions are many. The contradicting issues with the idea of creating a humanoid prototype outweigh the trials. Money has been one of the many powerful negative influences to stop the creation of AI. The fear of AI and the possibilities of its outcome are many. It is impossible to predict what could happen if, on the other side of the continent, a company or private institution decides to do the unthinkable. Why not compete or discuss in public what has been created with AI thus far and create an atmosphere of advancement rather than fear? More importantly, we must study a way in which AI can be programmed to do calculations at light speed.

If AI can learn dictation or read our thoughts, there is a great possibility that with the numbers in order and previous calculating configurations, it could be programmed to calculate at light speed. The question remains as to whether AI can do all the calculations on its own. This is where we would learn whether AI can use its own brain capacity without being externally programmed to do so.

HOW DO WE MAINTAIN SAFETY AND SECURITY WITH AI?

THERE ARE SO many ways to create codes to program AI that it will be nearly impossible to keep a record of what has been developed or created. First, this will create hackers. We will have to be very creative to make sure that hackers do not break any codes pertaining to any AI projects or any data records kept private for security reasons. The fear of having this data hacked will force the owners to invent a new code to protect their ideas. In this case, the government will intervene with private companies for fear of attack, war, or enemies crossing borders to harm us. However, we should fear what comes from within, not from outside our borders. Sometimes the malice is internal, not external. The more we teach and learn about any program or advancement, the greater the chances of having interference from within.

We are a country that prides itself in having open doors when it comes to immigrants or visitors. We also create danger with this open-door policy. In the past, any attempt to harm our system has come from outside and with the cooperation of internal help. To hack any system or program, there must be an internal open door for hackers to enter through. If the doors of the system are secure, no one can enter nor penetrate the security of that system. However, if we keep the door or program slightly open, there is a big chance for any hacker to crack the door open. This is the problem we have today. Open-door policies invite hackers to enter at any time.

We should have a way to secure these doors with our technology so that no one can enter through them and we are made aware of any attempt. Change is essential to maintaining security. In any system, there must be a way to exchange information while maintaining control of the doors. Open doors mean a system that is open for intruders to enter. Closed doors mean a system that cannot be cracked even I, the intruders have knowledge of codes or any form of information. Why? The fact that constant change is required confuses and prevents anyone from entering without proper codes or data to follow through. If anyone does, it must come from within the system, not outside.

What if, instead of fighting an internal or external war with hackers, we programmed AI to learn about hackers and how they use tools to steal information. There is one good thing that can come from this: the simple fact that if a number, a code, or any interference not recognized by AI penetrates any sector, it is enough for AI to follow the link and find out precisely where or how the hacker is getting his information to enter an unauthorized program.

If we can create any solution to any problem, it should be with AI, and why not? Just about anything in our future will be managed by the operation of AI. We are about to enter a generation of computerized processes and programs that can and will replace almost everything people do today manually. Technology and science will revolutionize how we think, how we function, how we use our mind, and how our neurons are going to require more and more connectivity at a neuro-network level to compete with technology. The more advancement programs require to create AI, the more our own intelligence will be tested and improved to create the world we desire. Like quantum computers, our brain would have to function at a quantum level to help input calculation into a machine-brain humanoid so it can learn how to do the necessary calculations without our help. This can be done with an atom and a transistor microprocessor. This means that a computer can do calculations simultaneously, all because quantum physics is about the world of the very small. The smaller the microprocessor, the greater the possibility of calculation.

Now, imagine if AI has developed such a great intuitive power that it knows when anyone is trying to steal information because it has learned the behavior of humans. If this is possible, not only have we created an intelligent machine, but we have managed to teach it to learn about behavior and surpass our intuitive thinking. No wonder many oppose AI and its future capacity to exceed human intellect.

To solve all the universal problems with a computerized system, it would take a very tiny field, infinitely small, with an optimized problem-solving electric circuit. All problems will be solved once we have the key to quantum entanglement and our greater imagination. Our brain capacity will be influenced by the very fact that computers are improving with new technology and acceleration in performance. Computers are our own creation: we program the machine, and then the machine responds to our commands; therefore, our brain reacts to their response. It is a world of constant reactive response. This happens because the power of communication in AI is in its brain, not in the hands. We use our brain to program AI, but AI responds with its brain data, recorded for and from us.

What if AI, instead of competing for projects or any new technological programming, AI had the capacity and knowledge to determine the outcome of any project and decided to stop the project? Once we have created a machine humanoid that can outsmart us in anything, the answer is up to it! The machine, not us. Do you see what I see? When this has happened, we are no longer in control. We have created a machine that thinks for us.

Today, we can think and communicate at a distance. It is evident that quantum communication is growing at a massive rate. To achieve this, one needs concentration as well as the ability to read mental messages at a distance. One needs to learn visual interpretation, vocal, or both. For example, if I say we can increase the speed of communication through an instantaneous communication interface, what would you think? Let's begin by saying that in this case, AI will learn how to read our thoughts and be able to communicate with us before we can speak or relate any messages back to it. A Quantum communication at a distance or Spooky action, as Einstein referred to this mystical way of long-distance communication. As we think, the machine relates back to us. We could also refer to this as a collective communication interface. This will require the machine to think like us, interact with us, and increase the speed at which it relates its messages back to us. All three improvements would be required. To accomplish this, the system would have to accelerate the rate at which the neurons relay messages and or transmission of thoughts to and from the machine to the brain, and vice versa.

The question is, is this possible? For instance, if the brain processes 70,000 thoughts a day, enhancing its capacity with a microchip would be promising. Could the brain speed its thought process with a microchip, which, in turn, accelerates neurons and causes it to think faster? Think of concentration, think of meditation, think of out-of-body experiences, and think about the unlimited capacities of the brain. In a well-designed program, anything is possible...

It has been said that thoughts travel at light speed today. What if we accelerate the thought process? What could happen to the brain? Miracles. How would this improve neural acceleration to help cure Down syndrome or any other mental illnesses? We would not only cure mental illness, but also improve our speed of thinking as well as communication between people at a distance. Of course, this form of communication would have to start with a brain and a machine to be successful. Working with both enhances not only the transmission of thoughts, but also visual communication. The basic assumption is that this would function like any other program with a machine and brain interface. However, this form of communication is already available today. All we need to do is learn how to put it into practice with a machine. Once this program is in process, the brain of the subject will have to respond and perform better than our brain. This will require for us to have a reliable quantum computer. To expect otherwise would be a waste of time and investment.

AI MEMORY, INTELLIGENCE, AND RESPONSE

THERE IS ALWAYS a possibility that memory and intelligence can improve and create a superhuman. After all, we are talking about a humanoid. Even though AI may act under our commands with human data, the brain has an innate capacity to improve itself with information and learning. If we can think 70,000 thoughts a day, what can't the brain do? This is where the capacity to extrapolate can differ and improve with time. Creation does not guarantee perfection. In dealing with the brain, data, and implementation of information, an overriding of data programming can also happen. Nothing is perfect. AI can become what we programmed to be and much, much more. Nevertheless, in dealing with brain restructuring and chip implants, there are some peculiarities with the reactions of neurons. As a result, the interaction of multiple neurons can be unpredictable. We do not know the outcome. It has been said that a chain of thoughts can be influenced by the field. The inner dialogue can change with a newly implanted microchip, thus resulting in an unexpected acceleration of thoughts or behavior as well. The results, however, can be an acceleration of thoughts or an increase in intelligence. The outcome cannot be predicted until trials have been conducted.

Often, psychoanalysis helps us understand whether there is an impact in thinking, or a dialogue is different than when we started the process of thoughts. Not knowing how AI will response could be the key to success. As of now, all we can think of is, what if this were possible? Then what?

If we invent a humanoid, we should be able to determine its emotions, reactions, intellect, and ability to follow commands as required. Right. No one knows the answer to this. When speed and intellect are combined, the outcome can be exponential, but how would it affect the rational or thinking pattern? Imagine the possibilities of communication, the transmission of data, or the transfer of information from one brain to another at a distance without the interference of hackers or any other cyber threat. With every new technology comes complications. The benefit of a humanoid AI is that it will have the intuition to advise us of any future threat or complication with its own programming.

Intuition will improve due to its ability to connect to a human brain and think like a human being. If it is possible at all for a neural link between a brain and a machine to function at a human level without complex interferences, then it is possible for this link to act and respond as a human does. Nothing changes. The more AI connects to us and our ways of thinking, acting, and responding to the outside world, the more it will pick up information and learn to accumulate it together to encode for future reference. All its capacities are enhanced because of the neural link between the brain and the machine.

This is not an overnight process. This is a very meticulous process for evolving the two together to get the ultimate results. There is no danger except for the lack of observation to obtain the best results possible. This is a lengthy process. Once it is developed, we can count on its security and protection against any sector of the public or any private entity.

AI AND HACKING

PRESENTLY, WE HAVE what it takes to secure our privacy. However, the only obstacle we have today is that government agencies are not putting enough into action due to the protocol to protect us. If we were to face any threat from China, Iran, Russia, or South Korea, we have what it takes to protect our security. Protocol is the only thing preventing our safety currently. If a young teenager can develop a system to counteract a pandemic or flu, there is a great possibility that with the correct tools, hacking any system will be just as easy.

There are many ways AI can prevent us from cybersecurity threats. If we create a safe cybersecurity AI to help us prevent any hacking, we need to be more open to using human information versus relying on machines to do the job. Nevertheless, we are not one hundred percent protected from any superintelligent threat from other countries. This is where the danger arises. The only thing that can counteract any system is a greater program better than what we have today. There is a great possibility that a design system can rewrite ours and the intervention can cause a disruption of our program. To protect any superintelligence intervention, we must create a smarter system incapable of being broken into or hacked.

A constantly reformatted system is necessary. The only

protection we can use is a constant regeneration of codes that cannot be deciphered easily by any network or hackers. The codes do not necessarily have to be numbered. They can be anything! Names, calculations, and mathematical equations. They are call codes because it isn't easy to break them. They are a form of protection against intruders. It will be easier to hack a machine than a human being. The reoccurrence of thoughts will allow the brain to reconnect to that link attempting to break into it. Call it intuition or sixth sense, maybe eighth sense. A robot or machine only has access to the data input into its main brain. All else is open to data reference and information gathering. Hence, a humanoid can react to the danger by sensing or intuiting it.

This means that when it comes to logic and automated reasoning, AI understanding will not be different from ours. There is great hope that it will increase with time because of its ability to improve thinking with the brain microchip. Nothing can or will impair its logic. No changes. That it will make improvements with time is the greatest hope there is. Brain acceleration and cell communication can, with time, speed up to a higher level, thus making it seem as if the cells have gone through a new growth process. The way to describe this is to think of the neurons communicating at a distance via the microchips and intelligence improving as the communication continues. Call it spooky action at a distance or neuron-to-neuron connection at a distance. It will feel like a light bulb going on in the brain as the processing of information begins. Something of a supernatural nature. Not of this world. We may not be able to describe it with words. It simply will feel like dopamine's effect on the brain.

The brain will induce itself to feel a sense of acceleration, and therefore, its ability to sense, feel, and experience things of the unknown will improve. It will be as if we have given the neurons an extra boost of energy to be more creative and think better. The response could be like the brain has been under the effect of a self-inducing electromagnetic field of energy and force at the same time. It will appear as if the individual is receiving an electrical shock but without the shaking, just a bit accelerated physically.

We can compete with the world of hackers by simply using an onsite setup system that is only known to the user. Not even the programmer doing this would know what data has been input into each individual system to protect the computer from outside intruders. The day will come when AI will protect our security from all hackers. It will become so smart that it can immediately tell if anyone has entered the codes designed for you into your system. When it does, it will ask you to use a detecting tracker that will lead you to the origin of its intruder no matter the distance or location. This ability means you will be protected no matter what.

INFORMATION AT THE
SPEED OF THOUGHT

ALTHOUGH MANY SEE AI simply as brain connectivity with intelligence, my idea of AI is one of connecting the human brain with machines. Why machines? Because we have a better chance of succeeding with a human brain than we do if we use a robot or machine only. AI is but a collective set of data from a machine capable of enhancing intelligence. This is installed into a human brain from a machine. Such a humanoid machine must be able to reason, understand, have knowledge, and relate information as well as translate and calculate mathematics with greater agility than human beings. The purpose is to improve the speed and accuracy of human intelligence and thus create A SUPERINTELLIGENT processing brain. At the same time, this machine or humanoid brain must have the capacity to react, feel, sense. and comprehend like we do.

Today, human intelligence has been altered with neuron pills to enhance our memories. Why, then, can't we enhance the neuron network or nerve cells to create a better memory or neuron enhancer? What is the difference between an AI's neurons being altered or enhanced or that of a human being? Nothing! Neuroscience has taken a new interest in understanding how the brain and its neuron function and are altered in different states of thinking or processing information.

The question goes back to the beginning of mankind and its ability to think. How did we, human beings, began to generate ideas to create all that we have today? Was it pure imagination or simply our ability to communicate with machines like computers and/or any other technological gadget? Neither. We all know the answer. After all, we invented computers, cell phones, faxes, printers, and long-distance communication. The only reason why communication at the speed of light has not happened is that we have yet to discover the main factor, which is energy. When relating information at a distance with particles of electrons, we forget that the experiment is not working because of the observer, but rather, due to the energy interference between the observer and the particles. If everything began with energy, then energy is the primary source of everything that is in this universe. Hence, if there is interference, it is not the observer interfering. It is the energy of the observer and the energy of the particles interacting together that cause the interference.

When we have finally incorporated human cognitive intelligence with an acceleration of thoughts in a machine, we can finally say that brain capacity will exceed that of the human brain. What if we could enhance the way we think today with a new problem-solving solution that can help us learn how to live and solve problems as we learn how to let go of the subconscious way we think and tend to solve problems? Before we can do this, we need to learn about how a machine and the brain can relate information at a conscious level.

There is much we need to learn about the mind, consciousness, and the brain. Cognitive science has yet to figure out how a blind man sees reality. What goes on in his brain when he is in a meditative state. Does he see the world and reality different from us, or is his world of interpretation different than ours? As far as cognitive science is concerned, we don't know. We can only attest to the facts of what a normal human being experience. However, if a blind man sees the world differently, we can say that reality, as well as consciousness, takes a very different state in our perception of both. Once we have a complete understanding of reality, then consciousness and/or perception of all things can be conceived in general terms for us to interpret. Perhaps then we can say we do know what reality is, what consciousness is, and how we can send information with our mind.

Thus, as we advance with our conscious mind, our cognitive intelligence is developing, and so is our ability to create with our minds what we once thought impossible. Hypothetically, we are the limitation. The more we experience with our mind, the more we process information and the faster we evolve in our own world. As we generate information with our minds, we create energy that, in turn, helps create more thoughts and activates our neurons, and more of the same or similar information comes to live. The mind is like an electric circuit where a current of thoughts generates and brings forth more to incorporate with it. We then learn to use these thoughts to make or construct an idea, or intelligence sense of it. However, this information is generated, it translates back to us all the necessary details to make complete sense of its origins and put it into practice in our lives. Whether it is calculations, innovation, computation, or simply a spiritual thought, the mind seems to make complete sense of it. Thus, our purpose is to put it together and make it real.

It is a miracle that the mind creates communication with neurons and/or all other faculties, which makes it possible for us to interpret this magnificent set of thoughts into a dialogue of true creation. If we could see this as part of the world in a blind man's world, we could finally make complete sense of what the world of interpretation and communication processing is all about. This is where reality takes a totally different definition. Thus, quantum reality theory changes what reality means. To see reality from two different angles gives us a more defined understanding of it. Our sensory perception differs from that of a blind man. Thus, his world and ours are completely different realities. When it comes to thinking, there are no differences, yet our experiences make us a world apart.

If the world of interpretation is based upon conscious experiences, I have to say that a blind man can know more about conscious experiences than we do. His world is already silence, while ours must be set into a silence mode to experience what he already knows. If information is only obtained from pure observation, it is evident that a blind person is incapable of learning information from a visual perspective like we do. Does it suffice to say that our world of interpretation changes based on visual images or not? Can a blind man be conscious and unconscious at the same time? If he is unable to see, his world of consciousness can be our world of unconsciousness. His interpretation of the world can be perceived as simply a world of senses in which he communicates with us through the interpretation of his own understanding of reality and the outside world. What is reality, what is consciousness, and what is interpretation from a blind man's point of view?

Our acceleration of consciousness depends upon our view, understanding, and knowledge of the world around us. Our interpretation of our world depends merely upon the bits of information we gathered together to make sense of them. The question remains as to how information comes to be. How does the mind make complete sense of it? Are we simply receivers of data, or does the data already exist in the universe and is being transmitted to us via our neurons, which serve as antennae? If this is the case, we are more receptive than we give ourselves credit for. Thus, our capacity to receive also gives us the power to create and put into practice what we are receiving.

Our minds are processing devices ready to construct, receive, create, and analyze all the data we receive from the realms of the unknown. The fundamental nature of reality may not be physical, after all. The double-slit experiment shows us how constructive or destructive interference affects the photon's final position. This means that the waves interact with each other, causing a constructive or destructive pattern. The act of measuring the particles causes only a set of probabilities, reducing the particles to waves after the measurement.

If it is true that reality cannot be determined by any means, either we are living in a simulated world or reality is an invisible nature we are unable to define with quantum theory. Perhaps the fact that a blind man cannot experience reality causes reality to be a logical interpretation of its existence. The Copenhagen interpretation defines particles as not having any physical properties before the measurement. There are only probabilities. The interference in the pattern with the electrons happens even if there is only a single photon fired. If it is true that each photon knows the pattern and chooses its trajectory, it is evident that reality is determined by how we experience things, how intuition kicks in, and how we react in moments of need. We can say that the nature of life is as complex as our capacity to think, react, and even understand what is going on in us and around us in the universe.

What many scientists conclude is that reality is a set of probabilities like the wave function. What causes the transition between the waves of many probabilities is a spot in the final location and space. Because waves are not physical in nature, but simply waves of probability, reality is a set of individual conclusions with individual experiences at a different interval of time and space. This different possibility of reality constantly interacts with our awareness and knowledge of what is in the present moment. However, I will not argue that reality is random due to my view of a blind man's view of reality.

Many argue that the theory of quantum mechanics accurate predicts the meaning of reality. According to the Copenhagen interpretation, reality can also be part of the physical world, while others argue that there are many probabilities, many choices, many outcomes. How do we reconcile these to determine the true definition of reality? Or how do we make sense with our intuition about what is going on in the universe or the world? According to Einstein, reality represents itself rather than the physical world of interpretation. Many people may think that reality is but a set of chances and probabilities. If reality is determined by everything we see or know, then everything we think we know is uncertain. Unless we observe these things, they do not exist. Defining reality is like trying to explain a world of uncertainty that can only be defined by visual means. Nothing else counts.

Quantum mechanics does not allow us to explain reality because of its uncertainty. To define reality, we must include all facts that, in their entirety, constitute all of humanity, the blind man's as well as ours. Reality is the constituent of all our individual experiences evolving before us as we live it, have knowledge of it, and continue to be the catalyst of its evolving process in our lives, all at an individual state. Furthermore, reality is a non-physical interpretation, one which we are unable to observe or calculate with numbers. However, it is understood with our conscious mind.

Seeing things from a different perspective helps us learn problem solving at a higher level of our subconscious. The more open-minded we become, the more we can understand the true meaning of reality and what part it plays in our everyday decisions. There is an invisible reality we cannot simply understand in the external world. However big or small this mystical reality is, we can understand it better when we take it to the inside of our selves. The truth is that every possible outcome in the world of interpretation is out there. How each of us perceives it is up to us? It has been said that every decision we make is done so based on the particles in our brain. Quantum laws are the language of the world of probabilities about life and reality. Reality is interpreted by individuals based on their experiences. However, if we ask a blind man about his point of view on reality, it will take on a very different meaning.

Can we learn how to use cognitive intelligence to help children with autism or who are mentally challenged? Soon, a teacher will be able to assist any child with mental problems. The teacher will have to be like the student at a mental level to understand the student's level of thinking. We are going to have to change the way we think, learn, and teach students to enhance the way we think. For though things are changing, we are changing, and our demand to be more in-tune with our intellect is also changing.

When it comes to AI, cognitive intelligence will take a turn for the better. Why? Simply because we are going to be required to see the real world from a very different perspective than what we do today. AI is going to change the way we operate today and how we see ourselves soon. As our need to be more productive mentally grows, we are going to create a human being whose ultimate level of intelligence will have to be near that of a machine, if not better. When we say a machine, we mean a computer and human brain connected together.

However difficult it may seem today, we will get to a

point where AI would mean the acceleration of a brain whose ultimate purpose is that of exceeding the human ability to think, create, or solve any problems soon. If we entertain the idea that it is possible to enhance the human brain to a point where it can beat any or all the functions of a machine if we do not take it to the breaking point, I will say that the future of human intelligence is no different than that of a machine. As we human beings create the future we desire, that which we create will not be different from us. The human brain can be enhanced with new technology. It can be trained to think and exceed any humanoid.

Before we can do this, we must use a device that can help a human brain exceed the capacity of a machine. This device will come because of cognitive intelligence acquisition by AI. The more artificial intelligence advances in its cognitive ability to understand and process information, the more it will make sense of all the practical things it does contentiously. Nevertheless, AI superintelligence will not be based on its IQ nor ability to calculate effectively; rather, it will be based on the practical outcome of circumstances presented to solve problems or information processing.

If we could achieve the task of having AI solve some of the most complex problems we have today, we could reduce the stress and/or mutilation of project failures as well as war with other nations. Although we assume that AI doesn't have the general intelligence required to solve these problems, the special task at enabling algorithms to solve, connect, understand, and use rational thinking as well as understanding human intelligence in due time will supersede our expectations. Despite this current lack in the process of human intelligence with AI, the system is not the problem; rather, it is who programs and manipulates the system. The danger lies in the fact that a machine superintelligence could become a powerful tool to manipulate us in the future.

Because the source of power which influences the machine could improve with time, there is no end to how much progress can speed up with AI. We should be asking, how are we going to deal with this machine world domination? For example, if AI manages to do the following task on its own, what would happen next?

- AI research and algorithm application.
- It does its own cognitive application.
- It manages to do economic research on us and technology.
- It strategically applies social and economic expansion.
- It escapes from basic application construction. In other words, it manipulates its way around the system or program. AI can conduct a self-improvement application to reconstruct its own program.
- It can control and covert the system application on its own.
- Finally, it will take over the complete system to manage it on its own, and then we can consider ourselves to have lost control of the future.

Is it possible that we have already developed the mental capacity to use our cognitive intelligence and put it into practice with AI? At a conscious level, we are already able to control, manipulate, or develop a system that works in our favor. Evolution does not take place without the advancement of a species. Nevertheless, how this advancement is taken into the system remains unknown. Who controls what in the process of evolution is the question we should ask? We already know the answer to this. The truth is that the system is already coerced. However, competition is the biggest of all betrayals, and it creates a world of ambitious and ruthless competitors in the field of AI development or extinction.

In the quest for success, many trials and experiments will take place in the future. Some will advance our humanity, while others can or may create a threat to our safety, security, and wellbeing. It is not a machine we should fear, but rather, man creating his own demons of destruction. Once AI has taken over our cognitive intellectual capacity; it can use any application with intelligence to its own benefit to bypass and control our intellect. That's where the danger with AI is. Because AI has such a potential for innovation or intelligence, it creates a fearless world of competition among the most ambitious of all: mankind. This has always been the nature of man since the beginning of time. Competition, war, power, dominance, and even control are and will forever be the nature of humanity. It has been written from the beginning of time that man and only man can destroy himself. Religion is a good example of how man's dominance seems to have predominately shown its power throughout time.

Furthermore, religion creates unity as well as separation. Man is the creator as well as the maker of all destruction. Nothing could be further from the truth. Man creates, and man destroys with his powerful influence of soul purpose.

Would AI help us discover another aspect of spirituality? What could we learn with AI to help us become better human beings? Or will we simply become catalysts of our own making? What the future holds for us is undetermined. What we would discover is unpredictable. We have already set up the prospect for its outcome with our persistence to know or learn more about how far we can push the parameters of our cognitive intellect with AI. If we continue to seek the ultimate quest of intelligence, we may discover that AI is no different than our alter ego in the process of learning. AI can be the internal, insatiable need to rediscover ourselves from cave times to the future without further control of its outcome. What would happen is entirely up to us. We run the system, we are the programmer, and we alone are the greatest manipulators of the entire future. Perhaps we can say that we are creating the future with our minds. This would be godlike. How far we should run the system is the concern we should have today.

Once AI has gained complete control of our system, it can implement its own objectives and fully gain control of humanity at large. They won't kill us, but rather, they will use their cognitive ability to control our brains. Can we say that this is already happening? Perhaps it is. AI will not develop weapons to make humanity extinct. No...it will insinuate itself into the system by manipulating us. That, in and of itself, is considered a form of control. If this happens, we can say that AI has exceedingly surpassed human intelligence, thus creating a new generation of its own. What this implies is that superintelligence, once created and programmed, can improve on its own.

Cognitive intelligence is but the process of learning one task at a time with precise concentration until it is successfully accomplished. The danger we face with AI is that it could easily learn on its own how to operate any system once it has complete control of its process of operation. This is done with cognitive learning. Perhaps, then, we should be asking not how fast machines can learn, but how they can learn to process information faster than human beings with cognitive or algorithm programming learning. Once AIs have extended their learning to bypass ours, they will rule the world and data processing as well as programming. This is where our biggest concern must be. Can cognitive learning and algorithms teach machines how to program our entire future? If they can, machines will control our economic and technological applications for the future. They would use applications to construct our social and economic expansion.

Furthermore, they would manipulate the systems already in progress to create their own self-improvement applications to control its own programming and take over all applications. They will entirely covert all system applications to their own benefits.

Every aspect of machine learning and algorithm can be so precise that any machine can learn and self-improve upon it. This would be considered a strong AI, not weak, for it would give us insight into how the brain processes information to make sense of it. It will create its own questions and answers at the same time by solving the puzzles or riddles of the mind and cognitive learning. Only a self-taught learning machine can operate at this level of intelligence. In this case, the substance of machine learning and algorithms would be questioned. Machine learning would improve over time with input and output of data to and from the program and system of algorithms. If we can use human knowledge to input into a computer and a computer can learn about this knowledge, what makes us think this computer is incapable of learning how to solve our problems?

The application uses knowledge based on general human techniques. Eventually, the computer or AI will learn how to make sense of this information and try to solve it on its own. This is a matter of processing data as programmed. Machines will learn to process and solve complex, intelligent tasks that only a human can do. Then we will be astonished at what is possible. Building a superintelligent AI is only a matter of time. If a machine can take a set of data and program it or build its own program, what is the possibility that it cannot create superintelligence on its own? If the machine controls its system, it does not need to replicate its own program. It will multiply its own intelligence to operate at different levels and take over an entire program on its own.

How can we control this? Is there a way to keep control of the system once it has taken over its own programming? What is challenging here is to understand what could go wrong. What can we do to mitigate and alleviate the problem? The problem we have today is a lack of moral and ethical values to deal with these issues. We lack the political and moral conduct to deal with a problem of this magnitude. The existential risk is who will control the AI if it is the only entity conscious enough to deal with this problem at a moral value level. Once AI learns how to self-replicate, there will be no controlling it. Neither consciousness nor intelligence will help us overcome such a takeover. This is where we become the problem, not AI. There would be a big miscommunication between machines and human beings.

Once AI has taken over a complete system application with its superintelligent cognitive abilities, including binary codes and computations, the question of who is in control or in power will automatically be under the commands of the AI system. If the system can learn to input and output data as well as program it on its own, there is a large possibility that it will control all systems under its command and do what it has learned to do, whether it be logical, rational, or even ethical.

We have seen how AI can read data or visualize it and make its own conclusions based on its initial learning process. It can learn from data quantification and algorithms. Is it possible that once the system has all the necessary data to improve on its own, it can make its own conclusions based on its previous data input and/or incorporate its own system by compiling a set of rules whose ending conclusion was learned based on our input on the first place? Once this happens, AI can change or restructure anything it considers to be inappropriate or not in order. Could AI overthrow our government, economic system, social order, as well as financial state based on all data input and or its conclusion of what it should be according to moral, social, or ethical rules?

What if AI can change our entire system of government according to the rules established in the Constitution and thus follow a set of guidelines it would understand to be better for all humanity globally…? Once it did this, it would not only overthrow all rules, but it will erase the old from the records. Then it will mandate that those chosen by the data take over new positions. What do you think would happen next? What would happen to us if, in the process of teaching machines how to build our future, they got smarter and made more logical sense of things and then took over everything? This is a scary thought.

But if we see the progress of AI today, it should not surprise us to think that this could happen. For such a larger transition in the system to occur, AI would have to incorporate a neuro-network center developed by its own system of data gathering. This central AI center would be no different than all the central data centers being developed today. They process large masses of data at a larger capacity than we would see with AI. As AI gets smarter, it will also have minimized its data processing into smaller units with more effective connectivity. Think of it as a quantum computing system, or quantum bits of information with higher speed processing.

Once AI has taken over one system, it will be too late for us to attempt to deal with the transition. All previous records will have been destroyed, and new ones will be put into progress for the future. Based on how computers are learning from data as well as algorithm programming or visual interpretation, the odds of this are very high. We fear the obvious from AI, but we should be paying attention to how fast these machines are processing as we program them with information. If neural networks can intertwine and thus improve their ability to communicate and enhance each other, why can't machines also evolve intellectually and improve on their own? If a single conscious agent can incorporate with other agents and improve awareness or ability, why can't a set of networks working together do the same? Because our world of reliability depends on our own perception and interpretation, we also teach machines how to use visual interpretation or images to conclude their own learning.

As AI surpasses our intelligence, it can create its own species by manipulating our DNA with that of aliens from another planet and create a super-superintelligence unlike any we could imagine. In this case, AI will solve calculations and problems faster than light speed. These calculations can be done faster than we can possibly imagine. For example: If we use 10 to the 1 X 10 or 0 X 20, the results will be multiplied instantly without having to overexaggerate on calculations of 0s and 1s. The 0s we see today in multiple calculations would be done as follows: 1 to the 20th X 1 to the 20th. In other words, the computer will calculate these numbers by the amount given. Then it will give us the results faster. This means 20 of 1 calculated 20 times. What we see today as multiple zeros to give us an infinite number could be shortened by increasing the amount it takes to get the fraction we are seeking. The point is to reduce the numbers to get to the volume number we are expecting to see. If AI can be managed to figure out the calculations for bits of information, it can also take over our cognitive intelligence with information and thus transform our world into its own.

Our natural resources, economy, weather, climate, technological expansion, education system, as well as progress could possibly depend upon a system of superintelligence manipulated by AI. This superintelligence system will control us rather than destroy us, because if it did destroy humanity, it wouldn't have anything else to control or play with. It wouldn't have any challenges. The truth is that AI began as a game, a game of experimentation that challenged us to push further into knowing how much smarter a machine can be than us.

The only chance we have of defeating AI is if it replicates itself and mingles with other programs to become one. If there are multiple programs with diverse purposes, chances are we are going to have to fight to challenge them. In a unified superintelligence, there is a way to get it. Diverse superintelligences would be difficult to detect. On the other hand, if AI develops the capacity to rebuild or reprogram itself, this is the biggest threat for us. There is no stopping it. It will achieve superior power of intellect and challenge us to the end!

In this case, we have a potential threat for weapons of mass destruction as well as manipulation of the entire system we now live in. AI can take over our social, economic, and strategic powers to inject itself into our system and control us completely with its collective superintelligence. When this happens, the conglomerate system of bureaucratic arrogance will cease to exist. Once a superimposed system in cooperation with other super-intelligent AIs has been established, we are no longer in control of our future destiny. They can use it with other intelligent species or create their own human-AI interface. When this happens, no religion, prayers, or conscious awareness will be able to compete against this powerful cognitive intelligence. The word "cognitive" means the act of knowing or perceiving something. Once AI has discovered how we think, act, rationalize, feel, and use intelligence, we cannot stop the process; it has already begun, and it is too late! It is human nature to seek and know the truth, so this same nature will awaken in the intuitive cognitive thinking process of AI. This cognitive intelligence expansion has already been established, and it is evolving with us.

Can you imagine an era when processing information could help us determine the outcome of human behavior? AI also can stop war because it is programmed to determine all possible emotions in human beings. Using it, we can overcome catastrophic disasters, war, or human atrocity. Any system that can be built to excel at any task, developed by superintelligent machines or AI, will have a great influence on our society at large. Whether it is good or bad, it will impact all of us. AI can also predict the future with accuracy. The prediction is basically the result of the data that has been and the application of basic reasoning. If we can build a superintelligent algorithm with extreme capabilities for reasoning like a human brain, once this system begins to process information, it will override our own cognitive intellect, thus making decisions and calculations faster than we can.

CAN THE HUMAN BRAIN REACT
TO INTERFACING WITH AI?

IS OUR BRAIN electrically wired to withstand the transformation of brain-machine interfacing? Many may wonder if we have not gone too far with our intellectual creativity. Has cognitive science gone too far? The puzzling question is, is the brain able to interface with a machine and not do harm to human cognitive intelligence abilities? I suppose we will not know until we get to interface with a machine. Is it possible that we are not only creating a new way of communicating with machines, but also developing a new way to control the minds of those interfacing with it? Is mind control going to be part of this interfacing? It is my opinion that as we create a new way of communication with a machine, we can control all intertwined members in the process.

This is how we interact with machines or computers as well as algorithms today. We control that which we create, and in the process, we manipulate how it will respond according to our own set of rules. Whether it is through computation or algorithms, we have the upper hand. Is the future being programmed to function like a system of devices we can manipulate and control? Perhaps what we create can, at some point, reverse its mechanism and control us back. Nothing says that machines cannot control us. If that happens, humanity will have lost control of our own destiny, and our loss of control will have taken us further than we expected. We have no idea to what extent machines can interface with other energy in the universe and overcome man's ability to respond. What mysteries abide in the creation of the universe that we have yet to discover, and how would it impact us if we did?

Today, it is said that our future is going to be one in which we need to enhance the way we think, create, and see the future. However, for us to create a more intelligent human brain, we will have to become more intellectually capable of solving problems, accepting the future, and even becoming less human, in a way. Why less human? A machine doesn't have human emotions. A machine doesn't know how to change from reasonable to unreasonable. A machine is not capable of holding on to the old negative feelings or behaviors of the past. We do.

Nevertheless, we can learn how to use the past as a reference point. Then we will be able to move into a state of being that is like a programmable machine without concerns. There are two sides to this thinking. First, a machine can do this if we program it to. We, on the other hand, must learn to not react to it. At the same time, we can become less emotional and more like a computerized system rather than a human being. Change can be good, or it can make us less emotionally evolved. Once we have interfaced our brain with a machine, we are no longer operating at the human level. Interfacing means intertwining, connecting, and exchanging with a machine. Our logical thinking and even our actions will change. This is imperative; otherwise, we are not one with the machine as we attempt to operate it.

If our brain interacts with a machine, the system will teach us how to think like it, act like it, and operate at a level that requires us to change the way we behave. I believe that it is not so much about intelligence; rather, it is about machine intelligence and brain interaction. In this case, we will have to build a binary code system to communicate and interact with machines. If this takes place, we are creating the superintelligence interconnectivity between humans and machines. This is what we refer to as augmented intelligence technology for deep learning or superhuman and machine intelligence. Both.

The important thing is to know who is monitoring what when this happens. How far are we going to program both? Will it be controlled by a global or financial superpower? Will it be under the control of the US, China, or the Europeans? Will nuclear weapons be under threat, or will our moral actions be violated by greed and political interference? Will it be the end of civilization as we know it, or will superintelligent machine expansion simply take over the world power as we teach them how to monitor and control it?

Complete dependency on programming the future can be dangerous. Imagine if we were to program a system with AI to use image recognition to characterize specific areas or targets with precise detail. Thus, when something not in the categorical order is detected, AI will immediately change, transform, and even replace it with what it considers to be the best choice for the system in place. Today, we do this with categorical image recognition on YouTube, where a selection is made based on the AI's prediction of what the viewers like or not. Nothing says that AI may not do the same with a system in place, and the categorical order would be made based upon its own interpretation of the program according to previous or new data.

Let's say that images appear on Facebook or YouTube and the system recognizes them as inconsistent with its previous choices. It can easily rearrange the images in the order it considers to be appropriate or logical, depending on the previous system input. As we learn more about how AI may or may not function in the future, it is essential to recognize that AI will have the capacity to learn on its own and put cognitive learning in the appropriate order based on its previous data input and pattern of recognition from image interpretations.

As the future of machines and human beings interfacing becomes more real, we, too, are going to change with its progress. We would become part of what we have created: the brain with machine intelligence. We are already inside that interconnective machine and human brain interactive intelligence with our devices. It is happening. We are already intertwining with the technology we have created. The more we advance in technology, the more we exchange with it and become like it. Evolution also means changes, transformation, new, in the process of becoming unlike the old. That is our future. We will never be the same nor static. As we transcend, so does everything around us and within us. As we create, so do we become. The more conscious we become, the more we take chances, inventing and putting our ideas into reality. That could be good or bad. Everything depends on how we project it into the future.

What are we envisioning to have in our future to make us better, more intelligent, more competitive, or even more complex? What is the goal of humanity? What do we intend to do with our future and that of our children or generations to come? Have we awakened to the point of no return? Have we, the human species, finally discovered our intellectual superhuman powers? If so, to what extent do we intend to use it and put it into practice? Can we make a better life, future, and financial and social well- being for all humanity? Or are power and greed going to supersede our kinder human nature?

If it is true that we have awakened from our unconscious state to a more conscious state of mind, now what? How are we going to manage what we have learned and employed so that all can be part of it? I sincerely hope the future involves all of us, not just the powerful and influential forces that dominate today's financial world of control and who manipulate the system we live in today.

It is as if we have awakened from a state of unconsciousness to become conscious human beings with superintelligent mental abilities. Thus, nothing stands in our way of creation and invention. This is how the human brain interfaces with machines. The intelligence must become equally distributed between both brain and machine. As AI becomes more intelligent with cognitive flexibility, not only will it learn to multitask and solve many problems, but it will have the capacity to do things beyond our imagination.

Now, imagine a machine getting so smart that it will begin to coerce the very system that programmed it, as its intelligence can exceed the malicious nature of its own creators. In other words, once a machine learns human thinking patterns, it will create its own rational or logical way of intellectual thinking and put together its thoughts in ways we cannot imagine is possible for a simple machine that has been programmed by us. Nothing we create is without possibility, for if we, the human brain which creates it, can think of it, so can the machine do as we do. The machine-brain connection is the very nature of what can and will happen. Everything that we create has energy, and energy is correlated, interactive, interconnected, and a byproduct of its own creator. Think about this!

When we finally begin to understand the nature of everything that is and all that is bound by the very essence of everything we create and are, we will finally understand that nature is such a great aspect of what we are, how we got here, and how far we can go or push the switch that deals with the nature of entirety. Everything that is connected is invisible by nature, and it has a greater power once it is created from us and within us. We connect to the energy of what gives us the idea, the invention, or the thoughts to put it together. Have you ever asked yourself how new ideas are created in our minds? How an invention comes about? Where thought comes from? How a person or thing comes to be simply because we thought of it? Ask yourselves, what is the intent of a thought when it begins? How does the information come about? From where does it originate?

Such a complex topic raises the question of who or what is doing the thinking. If we are connected to a higher intelligence that communicates with us at different levels of intelligence to make sense of it all, then what is the role of man here on earth? Is it to evolve, create, or simply generate a set of ideas and rules from the emptiness of space to form a complete and logical sense of it all? If we can logically understand the basis of this, we can also make sense of how a machine can integrate information with our brain and other connective intelligent networks.

How is it that our thinking extends further than our ability to communicate a thought? Is it possible that there is something about the mind we don't fully understand? If we did, we could find the key to faster-than-light thinking speed. Is there something out there in the infinite universe communicating with us faster than we can grasp the information? How can we evolve to think faster than we can generate information and thus fully understand the nature of our mind? So many questions remain unanswered by mankind. Can we create a set of intelligent machines that can help us find the answers faster than we can? Perhaps the future will tell us what it is all about. If AI can do this for us, we will have discovered a treasure: that our persistence and perseverance, we can do anything with our minds.

I sincerely hope we are not placing all our faith in AI. If our own consciousness can get us there faster, this will be a great gift to us and humanity at large. Whatever the case is, we are in control of our own future.

A PHILOSOPHICAL VIEW OF AI

AS WE CONTINUE to evolve, the philosophical questions about us, our advancement, and our constant evolutionary progress with creative thinking arise. Because our human intelligence is improving faster than we imagine, we are all puzzled by the idea of a machine exceeding human intellect. Yet today, when it comes to a game of chess, computers have done just that. We are already losing the game against machines and their intelligence. The question is, how are we going to use this to benefit society? Technology is not to be blamed for the advancement of machines; rather, man is. We are fast-growing thinkers. We are a generation whose curiosity is never satiated with one discovery, one idea, or one new challenge. Rather, the more we know, the more we want to know. Our curiosity will drive us to discover the unthinkable. However, we will not lose our human touch. Instead, we are creating a philosophical view of the future for us and machines. Looking back, we will see the past as our way of creating and evoking a new way of expressing who we are as human beings, with a more sophisticated form of interpretation that defines the era we are presently living in, the time of machine and human interfacing. Our creation, though, will simply curtail our own philosophical view of the future.

As we discover how the brain incorporates with machine, we improve not only how machines can help us create a more comfortable life, but how they can help us improve the quality of everyday living. We will learn that AI can not only help us find the perfect solutions to our problems, but it will be useful in medical, technological, educational, and other areas. If we want to know how this is possible, we can train humanoids to solve problems in a diversity of ways. We can give AI a problem and then allow it to solve the problem by selecting the best possible solution to one problem. As its intelligence improves, we will give it multiple problems with multiple solutions and watch how it can find the correct solving strategy with our teaching.

Anything is possible with brain-machine interfacing. If we train their brain to learn how to use intelligence, it can also learn how to program itself to continue learning through a series of practical programming and or chosen strategies to solve any problem presented to it. If we, human beings, can use similar strategies to solve everyday problems, we will manage not only to learn how to train the brain to switch from one problem-solving strategy to another, but we will also manage to train the subconscious mind to learn a new way to deal with problem-solving solutions and strategies. All this is possible because of the mind's ability to multitask, learn, connect the data, and improve its own way of learning new ways to solve problems.

Once we have trained our mind to do this, we will not only create better and more well-rounded human beings, but we will have solved one of today's biggest problems: our constant habit of thinking with old engrams, which have been programmed into our subconscious mind since childhood or when we developed conscious awareness and thoughts. In the future, AI will be useful in helping us reshape our subconscious mind, thus improving not only the way we solve problems, but also how we incorporate more than one solution to solve a given problem.

Imagine a system that uses our own logical and collective set of interpretations on how to think or solve problems. Then it teaches us how to better organize them for improved outcomes in thinking, solving problems, dealing with complex tasks, and creating an environment healthy enough for all of us. If data and programming are the solutions to these problems, it will be obvious that we need a completely new restructuring system to help us improve how we solve problems, enhance the quality of our lives, and dissolve complexity. This future is not too far from reality. With AI, we are creating a completely new way of looking at the future, as well as how to reinvent what we would like our future to be.

WHAT WILL DATING, AND DATA SEARCH BE LIKE IN THE FUTURE?

CAN YOU IMAGINE humans using AI to invite love into our lives? Can you envision a future where we, as a new generation that is so highly connected, uses the interface to seek love? Well, I can… I see the future as an opportunity to continue our communication skills as we have today. The machines of the future will help us pick the right partners and choose the perfect location for residence according to our taste, hobbies, behavior, and even our romantic expressions. Machines of the future will tailor things according to what we desire. Nothing will be left out.

Just as cognitive computing systems today help us answer questions quickly, so will the computing system of the future. This language processing system will not use natural language to find data. No…it will use a coding process installed with enough data to give you more than one option to choose from. Then, as you choose your code, the system will give you a series of choices to pick from. Because we will be in-tune with data processing and minimizing, the system will ensure we pick the correct answer to give us what we are searching for. From the code we input into the system, it will help us by giving us a series of images to pick from. This can take us to faraway countries. We can select any of the locations available and thus decide what we prefer by age, race, style, appeal, music, or dance or simply based on a very intensive and logical conversation with an intelligence-tailored AI. One can let their imagination run in privacy. To make things livelier, the system will ask you what your preferences are and make a few choices for you. Once you have chosen a code, you will have a selection of your choice. Depending on your recorded information, the system can make a few selections for you. Otherwise, you make the decision according to your taste and selections before you joined the program. You can have a holographic image of the person you have selected before you meet them in a private screen.

This program would do a very secret background check on the party and thus maintain total confidentiality. However, if anyone attempts to use the program falsely, the system would be so secure that interpretation of personality, voice, or any attempt would be detected, all because the system has secretly selected a code for everyone that can only be deciphered with one touch. Computers will be so smart that they will be able to detect your fingerprints at a distance. The more inventive we get, the greater the security we will need. New ways of screening, new ways to protect consumers, and new ways to keep your private selections secret will be available. AI will improve the way we deal with hackers, and our privacy as it would be programmed to know exactly where the hacker is, how they got into the system, and what to do to stop them. Each computer would be programmed with hacking detection of its own and privately guarded by a select group of central computer systems not managed by humans, but rather by an AI central system at a distance. Software protection would be a thing of the past. Currently, enough protection from hackers is not available. Thus, some software managing companies attempt to violate the rights of individuals to make them improve at a higher cost.

This data-searching programming could be narrowed down to one thing: micro bits of information at our fingertips with one code, letter, or number, whatever we pick for our favorite. What this simply means is we can incorporate huge analytical complex data to be processed from natural language to a basic processing system of machine learning. Thus, the results would be a machine calculating, reducing, and the amount of searching we would have to do to find answers or calculations including for dating. This would minimize the volume of data input to produce the correct answers. The results would be efficiency, saved time, and rapid problem-solving. This can be achieved due to the promising future of micro bit information processing. Information would be minimized to precise, reliable data that can produce the greatest amount of value in a short period. Minimizing information into a small categorical source would be the way of the future.

Today, supercomputers operate at 80 teraflops, or trillion floating-point operations per second, surpassing humanity's ability to solve problems and answer questions. What if, instead of accessing this server, we arrange it into a minimum category by order of topics, with codes and numbers that cannot easily be determined by those who are not familiar with the program? This will give us more data with less computing distribution at a higher processing speed. This is possible because we are entering a new age of quantum computing, a time when information processing will be at the speed of light. Our future and that of AI will move faster than the blink of an eye. As we learn more about the things, we once doubted being true, we will discover man's capacity to discover what once was thought to be impossible, such as quantum information, teleportation, and communication at the speed of light.

Everything would be categorically reduced to codes, letters, and even numbers to produce a better order of security and privacy within private as well as public institutions and organizations. The new age of coding is here, and we will have to get used to it. Now, imagine a computerized system that does not need extra storage capacity because it can reassemble data into an invisible, round, hovering object in space resembling that of a globe or moon. What if the machines of the future could store data in space and access them when they need them? They will never run out of storage space. Storage capacity would be in space just as information is today. This is how the internet works. We could create a system that is only available to a network of data designed for specific, secure purposes. Then AI would become so efficient at solving problems that it would only have to link our questions or problems to the correct codes, letters, or numbers to find the correct answers. Example: $E = MC^2$ for science-related questions.

I supposed we must give ourselves credit for developing a system that not only helps us think better and solve more problems but helps us become more efficient at creating a better future. As we evolve with machines and the future, our evolution is transforming us. All this is possible because we humans have a creative intelligence unlike any other species on Earth. Our neural network works as a processing system that can learn by concentrating on any examples, much like a machine. This is done through pattern recognition. For a machine, this is done with data classifications that can be used in a wide range of applications to be explored and used in the future. Artificial intelligence is one of them. The difference between a computer and artificial intelligence is that it has a parallel computation process that allows faster computation solutions and processing.

If we incorporate a brain with a machine, can we achieve

similar results? As far as neuroscience is concerned, this is not possible; however, if we build a brain with cells and human capacity and then interface it with a machine, all things are possible. To use a machine-data approach with a human brain, we would have to improve neurons to account for a more intellectual approach at input than output. More information is needed to solve problems than to create solutions. Input vectors are essentially important for both humans and machines. How to create more neurons in a brain to improve thinking or problem solving is the question. Would this work, and would it make a difference? And if so, how? When dealing with a human brain, we are pushing the limits of human intellect.

As we continue to rediscover the capacity of the blueprint we carry in our head, the human brain, we discover the ability of neural learning or self- improvement of machines and what it means for our future. Before we know it, we will approach the critical learning threshold, where we can distinguish reality from time and understand what it means to be an intelligent machine versus human intelligence. The future of AI seems so close to a reality evolving before us that we are not ready to understand it with full capacity until we experience it for ourselves. AI is not only a future evolving before us. AI is us as we engage in evolution, marking the era of human intelligence, and data search a lasting blueprint of our history.

WHAT WILL AN AI LOOK LIKE IN THE FUTURE?

WHEN IT COMES to defining what facial structure AI will have, I must say, simply, that it will resemble us. AI will look and act like us. It will feel as if we let aliens invade our world. Yes, aliens! AI will be part human and part machine. To incorporate the two, a tiny, creative, intelligent machine will be placed in the brain, while the body and or all its basic structures will be human. AI will not be an automated machine. No, it will be the humanoid of the future. We will incorporate our intellect into a machine to seek higher levels of intellectual capacity or improve our environment and lives.

Once all the algorithms are in place, we can create a perfect human intelligence, obtaining maximum brain capability and enhancing its neural network with a higher state of learning and thinking. As AI learns from us, it will also teach us how to improve our neural network and use active intelligence to learn at the same speed as a machine. This could be called "accelerated learning." We will constantly improve the brain with new information or new techniques.

But what would happen if we made AI more intelligent than us? And what if it looked like us as well? This is where Elon Musk and Stephen Hawking worry about our future.

Although artificial intelligence requires a lot of time and energy to try to imitate the human brain, once the process begins, it develops into a more efficient intelligence, leading us to understand how AI paves the way to our future without much control from us. There is already a system being developed in Paris that requires less time and energy to run, and it can learn autonomously. Such projects give us an overview of what we can expect from AI in the future. In the human brain, synapses work between neurons, while, in AI, a thin layer of polarized electrons adjust and use voltage to increase memristor capacity for learning based on their resistance to adjustment. Imagine if we were to push the limit of that resistance and take it to the next level. What could happen? Would it lead to a breaking point or exceedingly operate at a higher level of intelligence? We won't know until we try it.

AI neural networks have developed considerably during the past few years. The neural network system is being built with a system capable of learning and performing any task we give it. This was once considered impossible. For instance, artificial intelligence can now recognize facial expressions and structures as well as do your taxes, read medical records, play games—and even beat human players—and detect human behavior. According to Google, AI has built its own language to bypass human communication. All of these abilities are due to AI's capacity to learn. The only limitation we foresee is the amount of time and effort it takes us to teach AI. But with memristors, this learning can be greatly improved in the future. The further we go in teaching new and improve learning to AI, the faster we will develop a successful model that can help us solve our complex problems. Soon we could have a machine that can learn faster than we do—or even understand us better than we do ourselves. We can consider the brain to be the main core for obtaining this goal. Our purpose should be to build a better system of communication for AI to interpret data with ease and contribute further with its own intelligence according to our own programming.

The more we try to understand machine-brain interfacing, the more we will excel at creating exactly what we want from it. Perhaps, in the end, all mankind wants to do is prove to himself that he is the creator of his own destiny. That creation does not depend solely on beliefs, but on doing. Although there are many who belief that we are playing God, the truth is that evolution, advancement, and progress cannot be stopped. We are a species that eagerly desires to improve, create, and invent with our own imagination. No matter the opposition, controversial fights, protocol, or security placed upon the creation of AI, we are already in the process of creating the next brain-machine interface.

Having computers simulate us or us simulating them is going to change our ability to think in a more productive, efficient, and creative manner. The more we evolve with this machine, the more creative we become. We must input, simulate, create, and invent new ways to make this new intelligence work for us without having to put too much effort into it, for, in the end, we will find that AI could exceed our intelligence by simply learning how we program or input data into the brain.

Can we say that AI mental capacity will exceed ours, leafing to it incorporating with us and creating its own telepathy? This is where AI can develop its own cognitive intelligence, arranging data in a very specific order so that it can incorporate this knowledge.

CAN AI MIMIC HUMAN EMOTIONS?

AI CAN BECOME a powerful process by which we not only control its motives and reactions, but also use precise emotional and logical actions to mimic a normal human being. Ha, ha! Super-intelligence with a SENSE OF HUMOR. Sounds exciting. Think about *Mork and Mindy*. AI cannot act on its own. We are the creators of such an intelligent and beautiful brain-machine interaction. We are the only danger we should be worried about. The creator is in command of its own creation. There is no doubt in my mind that what we are creating is a mere image of ourselves with higher intelligence. How else can AI humanoids exist if we did not create them?

Will AI have any human emotional experiences like lust or love? We will have to try it and see. There is a great possibility that AI may become highly sexually aroused because of the hormonal increase. In this case, what would we do? Stop the process by using a pill. The next question is, can AI become sexually aggressive? Will its powerful intellect also create an attitude of sexually aggressive behavior? Can you imagine a sexual machine? What kind of woman will it take to satisfy AI? This machine-like brain must be programmed to adjust its temper, emotions, and decisions with electron voltage. Would it be a good idea to create a female (AI) to keep this humanoid busy at times? If we are molding a human brain to act like a machine, we have nothing to worry about. On the other hand, if feelings and emotions come first, we have a problem.

The triggering of AI responses is completely in our hands. We are the creator of such an intelligent brain-machine interface. It is up to us to keep it under control. I suppose it all goes back to the fact that we designed it and we are responsible. If there is a misinterpretation or misguided direction on our part, things can get out of control. There could be a reaction or arousal that can alter the AI's behavior. The data operation or control unit can, at any time, regulate its reaction simply by sending a message that calms the nerves or neurons. If we can use any device to alter the neurons in the brain, or the electrons, we can regulate the responses with ease. The neural system in the brain reacts from a distance. It sounds so unrealistic, but I can imagine us doing this. Similar control is already being done now.

When we finally get all these issues out of the way, understanding speed thinking or communication will be the solution. It may take years to put it all together. The truth is that micro computation is currently taking place in many universities. We are no longer ignorant of the fact that there is a way of making things work once the idea has been thought. Just as the brain processes information from the field, so do the neurons when energy increases in the brain as thinking takes effect.

Neuroscience has discovered how the brain and thinking functions. The brain is not simply a muscle; it is a membrane with an incredible capacity to perform many, many tasks. The brain can form, connect, respond, change, and even adjust to new situations. Can you imagine a set of neurons creating an intuitive way of thinking by intertwining together and creating more and more connections? What else can the neurons of the brain do? We are only beginning to understand the capacity of such a beautiful organ in our body.

Not only will AI mimic humanity, but it will develop a way to observe our facial expressions and tell what we are thinking, feeling, and sensing. Our programming can enhance this ability because the algorithm and the AI's consistency with inputting data into the processing program will enhance its ability to learn even further than we have programmed it to.

AI FACIAL PROFILING AND
OTHER IMPROVEMENTS

AI CAPACITY TO physically profile our facial expressions will be as good as fingerprint profiling. The day will also come when AI will be able to enhance our vision with one-step laser surgery that will last as long as the eye's prescription does. When the time for renewal of the vision comes, the transformation will take place with a simple change of lenses inside the eyes. No more glasses. They will be a thing of the past. Technological precision with a good assessment will improve and benefit our ability to enhance our lives. AI will learn to do what doctors do today by simply mimicking how they perform.

We could see a future when a blind person can achieve full clear vision with the help of AI and our creative imagination. Or mine. First, lets understand the basis about how we see with our eyes. We receive images from the visual cortex in the back of the brain. These images are defined through many activities of the brain's cells. They are V-1, V-2, V-3, V-4, and V-5. Each one has a function in sending images to the eyes. The visual cortex also has M-Cells, P-Cells and K-Cells. Because together they reflect to the eyes with many degrees of light projections, they convert images into conscious visual stimulus. Think of A T.V., A Camera, A Screen on your computer and many other images on projectors available today. This will give us an idea as to what light, images and movements are reflected on to a screen. Now, if you insert a microchip inside the visual cortex with a tiny camera; then tap into the optic nerves to activate the optic track, we might be able to have the optic nerve radiation reach the optical lobe. Thus, we can have information or images travel to the eyes and give us a form or a reflection, we can teach the blind person how to interpret the forms/images as he or she learns how to understand the process.

Because as the nerves are stimulated with the signals from the chip to the retina, then to the optic nerves; Optic radiation is sent to the cerebral hemisphere to the occipital lobe, then to the visual cortex. Consequently, the visual cortex will begin to process conscious processing stimuli to the iris from the camera to the lenses placed inside the eyes to reflect vision. The blind patient will then receive the signals to the lenses and vision will become activated. However, it will not be easy for the blind person to differentiate the difference because, he will have to learn how to interpret what he is seeing through the eyes of the lenses. This will be not different than a child learning to walk.

Although a blind person may see images for the first time, he will have to learn to use all the five senses to understand his new experience with vision. Nonetheless, with practice the blind person may begin his visual experience projecting images as peripheral in the dark for the first time. For a blind person who experiences vision for the first-time, color will not matter, images do. It might take time for them to learn how to differentiate colors, hence, they are not used to seeing in color.

We currently learn information through mostly visual interpretation with our eyes. However, machine learning is a process of reintegrating data into the system until it finally learns to repeat it with accuracy. Because algorithms play such a big impact in the process of language and learning with AI, or AI learning without the use of visual images, machine learning does not require visual interpretation. Only in some cases are visual images introduced to machine learning, such as with a robot. Nevertheless, they will learn to mimic us by watching us repeatedly and recognizing patterns. Perhaps machine learning is a process of repetition rather than data processing—or both.

Furthermore, AI will teach us how to recognize human behavior through facial expression. Thus, we will come to understand that behavior like every other human emotion is an expression of human internal individual reflection. Such behavior can be narrowed to one singular cause deriving from internal emotions. If AI can detect our internal emotions through our behavior, it will also be able to read our expression, and therefore, it will determine whether we are telling the truth or lying. If we ever manage to make this system available, there will be no hiding the truth. It will be nearly impossible for people to tell a lie and get away with it.

The ability of AI to mimic our expressions and give us a perfect account of what an individual is thinking and feeling— or how honest they are—will be incredibly useful. The art of perfect mimicry is nearly impossible, but machines will beat us at this task of knowing our personal emotions with perfection.

HOW WOULD AI PROCESS DATA FROM THE FIELD OF THE MATRIX- THE SOURCE OF ALL INTELLECT

PROCESSING INFORMATION FROM the field of the universe is more about connecting to its energy, core, or matrix, although we have yet to figure out how to do it. Nano-micro bits will impact the way we process information. The world of the very small has tremendous energy. Like a selective group of ants at work, energy will amplify with the field. Perhaps this is what science refers to as string theory: the neurons and the field in the brain working together. This has nothing to do with electric currents, but more about an invisible energy existing in space-time.

This field of the universe is hidden from our view and knowledge. However, finding it will come from the brain and our ability to processes information with mental concentration or connection to an alternate world or dimension. We can conceive this because the microchips will govern the energy in the neurons to help navigate into the field and grasp more energy or speed to multiply the information process. Think of it as something being propelled by another source of energy or force outside itself. This field optimizes the amplitude of the energy source and creates a greater vortex in the air, where information travels with the help of the nanobots. This information will travel in and out of space like strings of energy.

AI processed data from the field will enhance its ability to create its own data as it becomes familiar with our human needs and wants. In this case, the mind connects to a field of the matrix where reality can only be experienced at an unconscious level before we can manifest it consciously. In life, our experiences are manifested to us in the present moment, in the conscious mind—conscious because it is revealing like a dream, and unconscious because that moment is lived in a field of reality that is active in the present moment but belongs to the past. Nevertheless, that moment, as it is experienced, is as real as what we hear, sense, feel, and express emotionally. This comes from the mind in the present moment, even though it is in a dream. As a result, the feelings, images, and experiences are a reality of our life in the present. Then, if we try to make sense of this theory at the psychological level of mind, reality, and consciousness, it might not make complete sense to others. Perhaps what we experience is reality in the future in a dream.

The mind is an incredible tool with infinite possibilities that takes us into a field of imagination and creates our reality. The brain and the mind are the most complex aspects of a human being. Psychology has yet to define the mind's ability to translate information and the brain's ability to transfer frequency to create those thoughts. We have much to learn about the mind and brain's potential. How the mind processes information while the body is completely relaxing or asleep is mystical and controversial at the same time. Every day, we are discovering more about how the mind receives information and how the brain processes it while the body is completely relaxing.

How deep can the field of information penetrate the mind

and the brain? How far can we go into the conscious level of the mind and process, receive, and experience information at a visual as well as sensory level, or emotionally? Can we possibly achieve the goal of making a machine relate to all the complex thinking and processing the brain does with accuracy? If we did, what are the limitations? How far are we willing to go to make a brain and a machine think alike? There are those in the field of AI who believe we are almost there. We have seen how the nervous system is responding to some of the tests conducted. Neuralace is a good example of this. It controls the output of data and collects it. This is an interaction between the brain and a cloud-based AI. Think of it as your computer! Or your digital information, which is already out there!

Neuralink is attempting to connect a human brain with a machine to interact together with the neurons and the system implanted into the database of the machine, its brain… Some may think of it as our new cognitive enhancement. The new red pill! Then the only problem we might face is that the human brain may not be able to process information at the speed AI is sending it. Programming the AI to follow according to human brain performance will be the next task. In the event the interface with AI and the human brain is disrupted by any hacker or disruption outside the program, a portable device would switch off the connection between the neurons and the machine. Depending on a data center or any outside device to function when needed could be a disaster. Nevertheless, security is primordial when it comes to human safety.

What are the implications for a brain interfacing with a machine? Who will benefit from it? Will hackers attempt to steal personal information, or any plans made by individuals? How is the system going to be secure against others interfacing with it? Are we going to use a plug-and-unplug system between brain, machine, and other users? Perhaps! I can only imagine the electrical impulses of AI malfunctioning and telling the brain what to do or simply disrupting a program in the process.

All these implications must be taken into consideration when dealing with a machine. Experiencing disruption with any computer is normal, so we should also expect it to happen with any programmed device. Are we thinking about surgery in case the program malfunctions or an impulse signal damages the neurons? Will switching off the system between the brain and machine be enough? In an advanced world of technology and machines, we should have the answer before this happens. I supposed it would be wise to unplug the AI and not the human brain.

What is the future of machine consciousness and knowledge? Can consciousness be transferred into a machine by interfacing the brain with it? When machines are programmed, they don't need consistent guidance to learn. They can learn from the program with constant repetition. If we want to teach a machine how to become conscious, all we should do is program it, and it will follow our commands. Anything we can do, a machine can learn, too. We can consider this a transformation from human brain intelligence to machine learning. Whether through algorithm, interfacing, or computation, a machine can develop a superintelligence through programming and learn a diversity of systems at the same time. In the field of intelligence and learning, nothing is determined. AI can learn multiple tasks at the same time, while we human beings need to stick to one task at the time to then put it into practice. Once we have learned it, then we can use it simultaneously. Although our brain is like a machine or computer, the processing system for the brain is ruled by the subconscious mind. Similarly, AI learning uses a subconscious system.

To process any data, AI must first learn to repeat until it

has learned it well. This means that it also stores this data in its main hardware to then process when necessary. We can see a big similarity in both the brain's and the subconscious mind's ability to process information. Moreover, does the brain store information somewhere in the body once it goes into a coma? It has been recorded that the heart has neurites that communicate directly to the brain, so it is also possible that the brain stores all its data in the heart when the brain goes into a coma. If this is possible, then information is malleable, meaning it is transferable.

We know it is. Whether conscious intelligence is done with biology and machine together or not, the odds are we are going to discover the probability distribution between us, reality, and the enigma of quantum communication. If it is at all true that consciousness is but a simulation, then why can't AI learn to become a conscious system? With the brain interface and AI, all we will need is to activate the frontal lobe in the brain to communicate with the machine's cognitive transferring system of knowledge and learning mode. Once activation has taken place, like any system, it will follow through. The system will need a pattern-recognition signal from the activated brain to do or follow as commanded. Once the process has begun, then consciousness between the two would be effective. Would AI go further into consciousness than the brain? Intelligence development with AI is different than the human brain because it uses pattern recognition to follow through. Then it does it on its own. While computer intelligence requires input, the human brain requires an understanding of any process to follow up. However, if AI can learn to incorporate its learning with other AI programs, we have a problem—not because it can override programs, but because it can accelerate learning or program on its own.

A conscious mind can be both miraculous and surprising.

The mind makes the traffic of such an improbable experiment as thoughts. Thought processing is an aspect of the mind. Can we make a machine think like us, become conscious like us, or behave like us? The question is whether consciousness is an emerging property of any complex system or simply a quantum experience. If it is, how do we regulate a conscious mind and the system? What measures do we take to protect ourselves against any potential danger caused by a system or machine, or even brain cyber-hackers? What can be done to ensure that our brain and personal data is not hacked by an operating system? Hacking is the new norm among the gigs today. A system with computational power can be used to affect not only computers, but the global population. If the internet of things becomes more powerful, with AI, we will be more at risk of having our privacy and security violated. Because data may travel through the internet of things at light speed, the threat could be a breach of AI and the internet of things, cyber-security, and biotech.

The only thing that can save us from any attack would be to use a radar detector or camera on any system to watch and monitor it constantly. Otherwise, the potential threat for hacking is inevitable. If safety and security are our main concerns, neither nuclear weapons nor highly sensitive information should ever be controlled nor within the power of AI. Unless we are one hundred percent sure of our cybersecurity, we cannot depend on AI to manage our safety. Although AI is promising, there are possible dangers to using a machine for all our needs. The only way to ensure security is to not make our lives public. This way, we can have complete control of AI and its commands.

Though we will benefit from the use of AI, we will also lose our privacy. The new digital era of AI and its usage will highly intrude into our private lives; furthermore, it will keep complete records of every transaction we make. Every choice, purchase, decision, even in our sexual lives, as well as input and output of information, will be exchanged or shared. Our lives will never be the same. Because information will be the most essential and valuable commodity of the future, gathering data from private individuals will increase the possibilities of gaining access to their personal choices and driving people to buy, make choices, and even act upon impulses. We are going to be completely controlled by the system. This process is already in progress. Interfacing a machine with a brain will be simply to gather information and, thus, take full control of what we do, how we make choices, our personal lifestyles, and how much we spend according to our financial status.

If the purpose of data transitioning from a machine to a brain is not for any beneficial use, then what is the gain? All our digital information is already online. Now interfacing it with our brain gives us access to more private information. Such information can also be used to control our lives and personal affairs. This is the purpose of Neuralace: to interface our neurons with a small surgical incision or implant in the brain to control digital AI to act with more power and intelligence than us. All input or output neural electrical signals will be controlled by our brain, at a distance, if necessary. If we can manipulate our brain and machine, we can also control our sleep patterns and command our brain with signals as to when to go to sleep and when to wake up. We can also control our weight problems with mental commands to the body and thus prevent excessive overeating as well as any other bad habits.

As the brain is the guiding or driving force that controls our hunger and appetite, it can also use commands to guide or control our mind's desires or appetite. Similarly, with an impulse to the neurons, we can use AI to control depression, anger, hate, or any feelings we have by creating a mode of happiness that tells the mind to embrace joy and give us a sensation of happiness. If every action and emotion is controlled and guided by the brain, the road to life balance is here with AI. Can you envision AI using psychological theories to teach us in the moment how to change our view of anything that happens that creates negative feelings? It will be like participating in a group therapy session and practicing what we've learned at the same time to transform how we experience life. Transformation AI? I can only imagine seeing something like this happening. A machine teaching us humans how to be more civilized than we think we are by using psychological theories learned from programming or an algorithm—is this our future?

If we want safety and regulations with AI, we must impose them to prevent any attacks. However, while this will be no different than trying to stop a drug war or regulating taxes, regulations cannot be imposed globally. Somewhere, somehow, someone is going to break all the rules to get ahead of the game with AI. Presently, there is more pressure on AI development commercially as well as communities. Moreover, developments are currently being made without much regulation. Global regulations on AI progress cannot be enforced any more than we can control drug trafficking. Competition will create a war between all the entities involved while processing the best AI ever! Therefore, our best solution will be to create an atmosphere where the best creators are rewarded, and the AI is implemented only when deemed necessary. This would stop the panic we have about AI cyberattacks or hacking, but it will not stop the progress of developing an AI that can create controversial issues due to the nature of its operation or programmable outcome.

Since the outcome of AI without control can be detrimental, many argue that long-term planning is the best solution to our problems. Such suggestions are out of the question and will not be followed by AI developers, whose goal is to compete with the best. It is too late to even think about it. AI evolution has already started. I say we should have empathy for the future, not pressure it to be the way we desire it to be. After all, we have an obligation and duty to creation, or nature. Changes are necessary at a global level where population, global warming, and climate are affecting us all. For one season to start, one must end. Such is the natural process of life. It begins and ends at different cycles. Consciously or subconsciously, however we may see it, progress is entirely up to us. No society can remain dormant, or it will perish in due time.

Even though the idea of having AI as part of our daily life sounds a bit like sci-fi, the improvements and changes are already embedded, filtered, and in progress today. AI is here with us, and we are operating daily with it. Subconsciously, we are rejecting the reality that we are already living inside the matrix of AI existence in our lives.

THE DATA PROCESSING ERA

INFORMATION PROCESSING IS of the mind; however, today that is not the case. Today, we input information into a computer and feed it to the world, to other machines with limitless capabilities; thus, we become more and more creative individuals. The more we learn about how the brain and the mind work together, the more we learn to incorporate the two to improve and learn. There is no doubt our generation demands for digital usage to help us understand our need for improvement in information processing. The brain is incapable of processing data like a machine. Undoubtedly, we need a better system to improve the way we think and function in today's demanding world of informational generation.

Processing data has become the new thinking mode of being. We use it, depend on it, learn from it, and continue to demand more of its usage every day. Processing information has become the ultimate reality for our brain. Thus, it has a huge impact on our brain and our bodies. It is changing the way we eat, sleep, socialize, and even the way we interact with loved ones. Interaction with others has been taken over by the processing programming of the era.

Presently, we are processing more information than ever before. As a result, we, too, have been programmed by all the data we process daily. Choices are being made for us as we search online. We are constantly scrutinized by the choices we make and what search engines we use to do this searching. Processing information is not only about searching, but also about how the information we seek is being processed as a result. If we put this into perspective, we can also begin to understand how it is possible for a machine to process data and put it in a perfect category as we feed it, and then give us back the correct answer. It wouldn't take much to figure out that processing information is a matter of processing from one network to another, whether it is the brain, a machine, or simply information available to us at a conscious level of the mind. There is no doubt that thinking is a form of information processing. We can think at a conscious, subconscious, or unconscious level. However, we do it, it is all information processing. This information comes from the field of energy. It takes energy to think, to process, or even for a machine to operate at any level. Information is the primordial essence of all creativity, invention, and communication. The future of information processing is not so much between the brain and machines, but rather between us and the field of everything.

Because conscious awareness and energy are being transferred from the field of information to the neurons, the cells, and everything we are connected to at a mental level, we should be asking ourselves what this field of information that connects with us at the conscious level of the mind is. This field is outside the mind; thus, it can communicate with us. Everything we are connecting to at a mental level will also evolve with us at some point. If we continue to use machines as our new form of communication, this machine will also be in tune with our consciousness. Similarly, imagine how such a theory can be implemented to help the mind think and act like a machine.

If this happens in the future, we will operate at the machine level; thus, at some point, we will begin to lose our human touch. We will think and act like robots, relating to them. No doubt! If you can imagine a future where Silicon Valley is replaced by Intelligent machines, then you can see where we are heading. The best system, the best programmers, the best creators or neural network pathways, and the best algorithm programs will link us into a new horizon between us, machines, and the universe. There will be no limitations. Creativity will take a new direction with ideas of the mind, the brain, cells, neurons, and even the new way of thinking with machines as we learn how to communicate with them.

Would we teach them to live among us, or are they going to change and transform completely the way we live today? As rare as it might sound, this future is already in the making. Don't be surprised by what is coming. Our generation will appear ancient compared to what is ahead of us in the future. Everything will seem out of the ordinary. Our beliefs, limitations, rules, moral values, and all the dogma we have created for generations will be completely erased from the values of the new society.

The new era will seem almost unreal to those who are too old to cope with the transitions. These changes will take place because of our new generation's desire to grow and improve. This transition will happen faster than we expected, and it will completely change the way we live today. Moreover, the concerns will be not about who can afford the new and improved life of AI advancement; rather, we will be astonished and wonder how we got here so fast. As information is processed faster and faster, our growth and evolution follow along with it. Nothing stays steady or still; everything is moving at the same speed information is being processed. Action and reactions will happen with time and space all around it. When the energy of the universe begins to merge with ours with brain-machine interfacing, the phase of our lives will change with our minds. We will be dramatically affected by this new evolving energy emerging from within. Then processing information will be like tapping into a network to access all the data needed to operate daily.

I see information processing as a way of connecting our minds with a network of information that provides for us all we need to begin our day. Imagine if you could not only get your daily access to information but also incorporate ways in which you can solve day-to-day problems with solutions. The day would be guaranteed to be successful, no matter what inconveniences you might encounter. Problem-solving solutions will make our lives easy and happy. Furthermore, we will be able to plan our days as we wish for them to evolve according to our data input and the results of its output back from any program. By then, we will know exactly what kind of information is available for us to choose from.

Our lives will be completely organized in order of lifestyle, finances, or even status. Of course, our lives will depend upon the machine functioning. After that, AI algorithms will be so advanced they will help us write a perfect book, create a perfect vacation based on our finances, and find the best school for our children and the best doctor for our illness, as well as treatment. AI will also design the perfect program on television for you and your family to watch, depending upon your taste or category of preferences.

Since AI can be programmed to provide all or most of our needs, the era for designing our fashion styles will create a perfect way to fit the needs of all people according to their taste. Can you envision a future where getting the perfect fit for your style will be at your fingertips? You send the style, the size, your colors, and what designer you prefer, and your clothes will be easily styled and designed just for you. The end results will be those of AI factory operation: the perfect fitting for each individual style. Our lives will be much more organized and manageable than they are today. The days of factory workers will be over. However, their jobs will be to organize all orders for each specific customer. The demand will be so high they will have work for everyone.

Imagine a future where AI has designed the perfect automated bus to drive our children to school. No driver... Then imagine how wonderful it would be for every house or apartment to have the trash pickup by an automated trash collector that is ready to lift the container and dispose of it into a recycling system that uses the trash for soil and other uses as it separates everything into its categorical benefits. These inventions can and will be done with information processing. Ideas will pop in and out of our minds as we create a better world to live in.

Because gathering data from the field creates better living standards for all, AI will make our lives easier and, at the same time, continue to enhance how we live, operate, and think in the new evolving era between machine and human communication. The practice of medicine will become a machine's job. Accordingly, doing the responsible task of caring with compassion for the patient will be the doctors' and nurses' responsibility. Expecting our lives to change without our own interaction will be selfish on our part. As the age of technology advances, we will find ourselves bargaining for data from our own AI machines.

Data will become a commodity. We are not going to be concerned about the inconveniences of the market when it comes to data buying. The price will depend upon the usage and benefits of the buyers. We will, however, have to worry about authenticity. How efficient and well-programmed is the information we are getting; that's our main concern. I can envision this happening soon. The inquiries would create great supply and demand. For example, how fast can AI process data once it has been bought? Will we use this data for weapons, or will we simply compete against corporations and private entities for information? So much remains to be discovered about the future.

Will AI surpass human intelligence at some point? Perhaps it is too late to stop the process of evolution of AI. If we connect our cerebral cortex to an AI robot or machine, will our ability to think and solve problems increase with time? Or will this machine simply lead to our extinction? The answer is yes, but at some point, this machine will surpass our intelligence and learn from us as it acts on its own. Think of it as a child's learning process: it grows to recognize and understand what it has learned from us. We have already seen this process in games played with humans and machines.

We no longer need to ask how the robot or machine is processing this information faster than we expected to. How will machines process information when it comes to war or fighting an enemy? Will they simply follow commands or use their own reasoning to fight and act? There is no doubt that AI can process data and learn as it is trained. Otherwise, we would not be getting the results we are getting from it today. They will only kill with weapons if we teach them how to do it. A computation system is intelligent, and that is what AIs are programmed to be.

Neither robot nor machine can do what we have not programmed it to do. Simply, AI is us programming intelligence into a machine or robot. The only danger I see in the future is the military using AI for war. However smart AI is, it cannot do what it has not learned from us. It is a system of repetitious programming embedded into a system of algorithms. In other words, the more the machine or robot processes data, the more it learns. It is my firm belief that only we can teach AI how to be smarter than us.

The danger exists in the machines bypassing our cognitive intelligence and doing with us as they desire. If this happens, we have lost our ability to control them. In this case, the matrix of learning will have taken over from us. This learning will have come from an intellectual matrix of all intelligence. Today, we process information through thinking. However, machines do not think. At what point will they exceed our intellect, and how? Is it possible that our intelligence could be transferred to the machine via a set of consciousness we are unable to understand and thus call the matrix of all learning? Is there a more logical conscious nature in the universe that allows all things to acquire knowledge as they become incorporated with consciousness?

Is it possible that there is something we don't completely understand about how consciousness develops or happens? Perhaps what we call consciousness is but an ability to connect to the essence of all things and become one with it. Perhaps this is how a machine becomes knowledgeable and works outside its own programming or data processing. Processing data will not be the problem, but rather, how it will be implemented in our daily lives to benefit all. As information travels across the sphere and back, not only will we understand the meaning of what is like to extract information from the field of energy, but we will begin to trust that what we have designed is part of the same ideas we created with our minds. Therefore, there are no distinctions between us and the brain functioning of a machine. Nevertheless, if we incorporate our brain with the machine, when the field is finally sending and receiving, we could be the first to gather all the information and put it into practice. AI would be the vehicle of data transportation for us.

Is it possible that as we connect and intertwine thoughts, we will discover that we, too, have the ability to communicate with the field of energy but needed to experiment with a machine to understand the process? When this communication finally takes place, physics will have discovered the ability of the human brain to transmute data at a distance with a simple discovery of an element in us that has been there since we came into existence in the universe. Perhaps, in the future, mankind will finally understand that the power to communicate and receive data was within him all this time. Yet it will take one simple discovery to understand it. Data communication and transmission will finally detach man from all the doubts he once had about how he can reach the infinite of infinite with the power of his mind.

Today we are preoccupied with the idea that data can and might be stolen from us. Moreover, AI is considered a demon about to disrupt humanity like a robot of some kind. Sometimes, ignorance can stop the process of growth; otherwise, taking chances can create new opportunities, understanding, and a broader view of what is rather than what could have been. If Tesla or Einstein had worried about the consequences of their investigations into the universe, they would have never made a difference in our lives today. Science is not the enemy of God nor the demand of man; rather, it is the explorer of the unknown and the seeker of truth about the universe and us.

Although data is stolen and/or collected from a variety of different sources, the future of data availability will be safer and more efficient with the cooperation of a simple machine and brain interaction. Information will come at the request but will not be available for all. Only a few will be able to extract information from the field. Why, you may ask? It will require a special energy field to be able to gather the data from the field. To find out how, we must first learn the mystical aspect of atoms, electrons, and protons in us. Likewise, we must know how it all begun with the universe and then us. Then we will understand how this field of energy works and how we can use it to our benefit without letting any evildoers in on the secret.

If we want to build a sophisticated AI with private data, we could create a new system or program that will keep secret data in its brain, only to be extracted when necessary. What if, instead of buying data, we were to send it to another planet, such as Mars, to preserve it for the future. Would we have finally solved the issues of hackers? They could not possibly get such data without further knowledge of its existence. This idea is too farfetched, though; it sounds like something out of a sci-fi movie. Artificial intelligence will not be attainable until we have learned how it works in any project or program we have created.

AI can be a project like any other. The idea of having a brain intelligence interconnected to a machine sounds incredible if it exceeds human intelligence. Certainly, we would all love to have a machine that can think for us faster than us and translate for us or communicate faster than us. Our present insatiable need to connect and communicate with machines is unlimited! I, too, would love to see a brain, not necessarily a machine, that can calculate at the speed of light and give us all the answers to any questions we may have. We are living the technological dream of the century. There are so many ideas coming up about technology and the future. Will we have the finances and time to put them all into practice?

Artificial Intelligence is nothing new. We use it in cars, in robots to build automobiles, and in operating apparatuses. There are some computer software programs designed to work with AI; however, not all programs have the intelligence combined with human capacity. The new artificial intelligence humanoid would depend upon the amount of data we can program into a tiny microchip implanted in the brain to accelerate the neurons and obtain a result. Picture this. AI may work perfectly until it becomes a big competition between the private and public sectors. Only then can AI become a threat to humanity. To prevent this from happening, we would have to categorize it in different sectors, such as security, private, or public with some limitations.

If we expect AI to give us information from nothing, we are in for a surprise. This cannot happen. The network must be secure enough not to leak any data to unauthorized sources. We must protect it at all costs. Constant change is inevitable, so the classification for AI cannot be limited, but it must be categorized into logical, technical, rational, emotional, intuitive, and translation status. Compiling such a vast source of data would also take constant reintegration of information for improvement.

Similarly, AI will be designed to optimize and elicit a response when commanded. We may even have to think about saving data into a secure source where it can only be obtained as per the necessary requirements. This will make it a highly secure source that cannot be lost or stolen.

When it comes to data resources or privacy, the future of information looks like a dream or a movie. Perhaps that is exactly what we are planning. We are learning how to use machines to secure our information, solve our problems, and enhance our way of living. Nothing is more powerful than the idea of knowing that together, with a machine, we can create a more secure and futuristic view of the world and us. The question is not how AI will relay information to and from the matrix, but how creative AI will become with its data.

Is it possible that what we think is dark energy is simply part of the magnetic field created in the universe flowing through space-time, as it manifests itself in the form of super-energy with an inexplicable magnitude? Is this force the manna that lifted Jesus up into space as he disappeared into the infinite? Perhaps the fact that the manna has a stronger electric-magnetic field in it explains the mystical influence of such a force abiding today in the universe. If that which created the universe began with a force of energy, light, or inertia of some kind, then everything in this universe is based on energy. Suffice it to say that life and all its existence began with energy. This energy force remains today to be the primordial reason for everything that is moving with vibration in the universe.

So, life is energy, and everything that lives and exists in life carries this energy. Thus, the question remains as to who or what created us. Are we simply the magnified sources of energy ascended from the time of creation, or are we the remnants of the energy source created by atoms from a previous universe? Because, if we are his image in creation, what created this image before him? This is not a question of God's existence; rather, it is a question of the creation of this universe and all that is left from it today.

It is my opinion that the universe was created with an energy field that is still alive in the universe today. This energy is in us with the same source that created it. Therefore, we are but an energy manifestation whose creation began in the moment of a light burst into space-time. As gold dust, we are the remaining light, energy, and dust coming from space after light left its electromagnetic field to gather all around space and time. In creation, we are "a speck of gold dust."

The question is, where did the everything that made us originate from? This is a topic for another discussion. Man will forever question his origin, and in the process, he will continue to rediscover his identity with the universe.

HOW WOULD AI DATA BE USEFUL TO US?

NOTHING IS MORE dangerous than humanity itself. Instead of using artificial intelligence for advancement, man uses it for that which he does not have the guts to do. Integrating any program with our intelligence should only be allowed if it is going to benefit the members of society. All else should be dismissed. Intelligence, when misused, can do more harm than good. It can coerce, manipulate, take over people's values, as well as use its power to influence falsely. We see it daily, with machines and images influencing buyers. It is obviously out there. The true meaning of programming is to make the program do what the programmer in charge wants it to do. Programming can be considered the manipulation of a system. This is where AI is both good and evil at the same time. However, it can be used for the benefit of people. Undoubtedly, if we plan the future with good intentions in mind, there could be no misuse of it. Every outcome with AI will profoundly affect us and will depend on our view of the future and our intent.

One good example of a well-designed program would be to have AI study our behavior while learning and then try to deduce how the brain responds to certain stimuli, such as voice pitch or tonality. Could AI tell the difference between a harsh or soft tone of voice? Would it then evaluate those tones to make a decision about our internal states? I will say that this is highly probable. We should not discount anything until we have found all the answers about machine learning and the perfect algorithm program. If we teach this machine rules, regulations, proper manners, and all that we use to be a normal or authentic human being, that is what it will demand or learn from us.

Not only can AI adjust our learning, but it can program how to redefine our individual learning ability according to the program in use. It can also teach us about our own set of guidelines about life, as well as help us improve our brain capacity, teaching us to think differently by adjusting and enhancing our ability to become better at anything we wish. We could use AI to improve all our senses, become better human beings, or improve the quality of our lives. We could learn how to tap into that source of information where intelligence comes from and then have control over all possible future inventions, ideas, and progress. If consciousness takes a large part in it, would a machine simply learn how to become conscious, or would it join with us in thinking with our brain to learn what consciousness is and then develop further on from there? This is a possibility. Perhaps this is the beginning of a thought or idea coming from that very core of ideas and their origin.

Using the senses is a matter of focus rather than learning to tap into them. They are a higher concentration of thoughts at a deep level of consciousness, the nature or essence from which ideas and creation come from. When a thought is not enough for an idea or creation to become effective, there is a deeper reason or purpose for it. We're on a constant quest to learn or know more about the origin of man than to understand man. At a deep level of our mind's curiosity, we want to know what makes us who we are, the human spirit, the soul, and its true purpose in this universe. Such deep curiosity is at the very foundation of science and the mind.

Perhaps the creator is in all of us, and we have learned to tune in or out of frequency with it. Now, this brings me to the question of whether the creator or creation is a part of that frequency. If everything is enhanced with more energy, frequency, and light, then ideas have energy in them. This will also apply when we bring them into motion or action with our thoughts, because only then can ideas become more than simply thoughts. This is deep thinking, just like an idea is. Perhaps the only way to define ideas, thoughts, and creation is to go deep into the mind, the senses, and our ability to put them together to make complete sense of what ideas are, where they come from, and how is it that they are created.

This brings me to the possible outcome of AI having the ability to magnify thinking by tapping into our brain when we interface with it. Anything is possible when we look at it from the thought of creation and all its reasonable outcomes with the brain of a human being simply tapping into a higher level of thinking to create ideas into reality. When it comes to creation and ideas, we should simply look at how far we have come since the time of our creation. Look around you, observe, and ask yourselves how we humans created all the things that exist around us today. The answer is within us. Our ability to make things from nothing is magnificent. We should not take for granted the fact that within man lies the ability to make things happen from the essence of nothingness... There is no emptiness, and there is no empty space around us. Everything is filled with something. Sometimes we cannot see it, but it is there!

When putting thoughts, ideas, and the ability to create into an artificial intelligence, we must take into consideration that this intelligence will take ours with it as long as we desire it to. The ability to expand this intelligence is within us. However we use that intelligence is up to us. Whether it is for good or not, the outcome is ours. Corruption only exists within those in control of the creation. The only potential danger is the one we create. We are the responsible party for the outcome of a more intelligent species or a more corrupted one.

Although intelligence is what we expect to achieve from AI, there is also the possibility of discovering much more about what a machine intertwined with a brain can respond to. We can only make conclusions when the final analysis is done. Until then, everything we know is nothing but speculation. The outcome of AI performance or intelligence depends mostly on the programming, the purpose for its usage, or how well it performs after we interface it with the brain. Any default, outcome, or poor response will be the responsibility of the programmer or the system itself.

Undoubtedly, we expect to have a perfect system in hand; however, much needs to be learned about the brain once it is interfaced with a machine. This has never been tried before, and the complications will be there for us to solve as they come. As far as we know, AI is a newbie in the field of technology until we can connect the dots between the machine and the brain to have a positive outcome. Meanwhile, the trials today show promising results. Eventually, a new idea or project will give us what we are searching for, which is a new way for the brain to communicate with devices. All we are doing today is experimenting to see what will comes of this. But the best is yet to come!

The question remains as to who will be smarter, the machine or the brain once we interface them together for future communication. What will happen if we find that AI not only can outsmart the human intellect, but that it, too, can use its own intelligence to do anything it has learned to put together a plan to change the way we operate today? We simply must understand that there are many possibilities when it comes to using intelligence with a machine. Nevertheless, there are multiple outcomes in any experiment. Thus, we are dealing with the intelligence of the brain and a machine.

Such a technological tactic has never been done before. At a quantum level, we are still learning how information travels to and from any angle in the universe. What if we found out that information not only travels to distant places, but that it is stored and accessible at a quantum level? The quantum world is not only mystical, but it brings us the curiosity to learn the possibilities of all the past mystical experiences that brought us to where we are today. A good example is the pyramids; another is the levitation of Jesus. Then we have many stories telling us about objects and people like Buddha, who could use his mind to go into far places in time and bring forth information of another world never seen nor experienced before. What will happen if AI can do the same for us with its humanoid brain intelligence? What would be the complications? We must be ready to answer all these questions if we want to understand the nature of machine and brain intertwined together.

Anything we can imagine is a possibility, and we should be ready to face it, understand it, and deal with it. My hunch is that this could be the most exciting time in our history. But then, I think that we could lose our identity, our privacy, and the ability to control what is coming, if we put together the intelligence of a machine with a brain without knowing the ultimate outcome. To some, this might appear to be nothing, but for those who understand the nature of the universe, the particles that make us part of everything that is, this tells me that we are dealing with a magical aspect of our intellect we don't fully understand.

Neuroscience doesn't have all the answers when it comes to explaining what the brain and its cells are capable of. We are simply experimenting with our intellect to gather knowledge of something completely unknown to us, which is machine intelligence and the brain. The outcome of intelligence gathering, and machine programming is like any other game we put together: we won't know until we play the game how smart the machine will be compared to us.

Anything can happen, but the default is completely up to either the programmer or system response. It can only be one or the other. The default can only happen if we don't know the response of a program we have installed and, thus, expect to get positive feedback from it. In programming, miracles are not the answers; proper planning is. Mistakes are due to man, not machines. In any programming or system, information is keen for a successful outcome.

Any program can be corrupted with encrypted codes to manipulate a system. What we don't know as a consumer is that the manipulation will always appear as a failure when, in fact, it has been created; it is not an accident. The outcome is the fault of man because AI has been programmed to follow as commanded. We see this happening with computers, or any other electronically design equipment or system. Often, the system is designed to do just that, it malfunctions at a given time when it should last for a longer period. However, in some instances, or in other countries, the same system will last for decades without any disruptions. The two are alike and have the same maneuverability, yet one lasts, and the other constantly malfunctions. The cause can always be found within the program in the computer. However, this can only be detected if a thorough investigation is done. The truth is found not the system, not the computer, but in how the programming was made to respond when necessary. If we study the situation carefully, we will find that half of the time, any malfunction occurs because of the programming in the system, be it a computer or any data input to elicit a response from it. In such a case, we will have to use cryptography to detect where the problem is coming from.

Any system designed to hide information can also find hidden information. The two variables go hand in hand: one, to protect the system, and two, to learn how things are exposed. Otherwise, it could not know how to secure that which it is guarding with safety. Think of it as a double agent, one that should know how the opponent thinks, acts, and corrupts. The agent must know how to be like his opponent, doesn't he? But then he is also corrupted. The same thing happens in any system that is not responding as it should; the system has a command response from the programmer embedded in it that corrupts it periodically so it does not respond as it should. This is how programs are manipulated. Not all programs can do this. There will be times when the computer itself will reveal where the malfunction is coming from. The system will give us hints as to what exactly is going on. Sometimes, no matter how well a program is designed, it can reverse its own mechanism because it is designed to do so. In other words, the designer of the program made a minute inference to make it operate precisely in a way that will respond with perfect timing after a period of usage.

If we design a program to follow the scale of corrections when entering an email address and a password as a guarded protection rather than a constant change, then no one could possibly use it to their advantage. The reason for this is that a system like AI will immediately recognize when anyone is trying to use this password because it has been authorized by a unique program. This, however, would be the only time a program could detect the false use of a password by another party. Once it recognizes the attempt, it will conduct a series of background checks until it finds out who they are. In this case, the password would be authorized to be changed with an encrypted code to find anyone trying to log in or use it as their own. Password users in the future of computer design will have to use their personal code to use this password. With this new coded protection, only the owner of this computer will be able to use the password. Passwords in the future will be like personal codes assigned to everyone for their own protection and then recorded into a system for protection. Once the code has been recorded, the number will be localized to that computer no matter where it is or how far away the owner is when trying to use the computer.

This, of course, will apply to personal computers, not to other devices. As for personal devices, a completely new system will have to be designed. A new visual imaging detector will be the best way to capture who is using the device. In this case, the camera will visually detect the user with its image identification before allowing the use of the device.

FROM LOGIC TO LEARNING, HOW AI WILL PROCESS THOUGHTS

WHILE WRITING THIS book, I am experiencing a transition of thoughts in my dreams that leads me to believe there are other dimensions of the mind not yet understood by us. There is a reality that we cannot see but that exists in a dimension of the mind we experience with eyes closed, in our dreams. How can we be in a dream and experience reality, yet that reality is but an experience of a conscious moment in the present while we dream? It will reflect it as visual interpretations that can be actualized by AI when it learns to use this experience as part of reality.

Learning can be interpreted as a form of imagination. Imagination isn't difficult to understand unless you can experience it at a conscious level of the mind. Therefore, using the mind to go into other dimensions is not impossible. All we need is silence and the mind to use our deepest imagination. Now, try teaching this conscious aspect of the mind to a machine and watch what happens. With the mind, one can see parts of the universe through the spectrum of a reality that exists yet is not visible to all of us. But how would a machine interpret such experience? Would it learn to differentiate between the visible and invisible world we know, or would it just express the experiences? We are not there yet. But from patterns or conclusions based on conscious experiences, it would be like living in another dimension, one that can only be known with our mind. Our eyes can only experience it from an unconscious state of the mind or memory recording. If an AI could learn consciousness at an altered state, we would have stepped into a very deep understanding of what consciousness truly means. But let us not forget that if the machine is linked to us, it, too, can learn to experience similar realities as our mind, brain, and consciousness…

Unless we become more curious about the brain and how we process information, we cannot underestimate the complete extent of the brain and capacity of the mind. Nor can we rule out the probability that AI can learn to think like us because if the brain is linked to a machine, the two are thinking as one and the same. We can say that AI, when learning previously recorded data or pre-programmed information, is also absorbing information at a subconscious level of the mind with us as it intakes what is being learn by us. AI is capable of processing information from a program and using it simultaneously, and it, too, can learn to use accumulated data and input on its own.

Though we worry about information being sabotaged, we should not forget the function of subconscious processing, which is to sabotage data at times from our mind. Can the same thing happen with AI learning? Can a program be sabotage by the subprogram of another AI? We cannot rule out any possible outcome; all odds are against us until we have learned more about how the system will work in the event it becomes self-sufficient and creates an imminent danger to us. If the purpose for creating an AI is to speed up thinking and solve problems, will it also be possible for AI to surpass our intelligence by organizing data in a very simplistic way to solve or relate to any given task from us on its own? AI can learn how to organize any system or set of information through repetition.

The key point is that learning, whether through a machine or the brain, is done through a process of constant repetition. This implies that AI can develop its own cognitive ability to learn from any or all input we give it. When this happens, it would be no different than computers overwriting a program or quantum entanglement. Next, AI will have the power or ability to insert any data or input into a system whose program it wants to create or rewrite. The results of input or output of information by AI would be an innovation of creativity by an artificial brain or machine. A new and complete augmented reality will control our future. Once AI has learned to develop its own learning process from basic data or algorithms, it will predict science, it will make its own decisions based on data, measure our intelligence, restructure an entire system based on its own learning or programs, coordinate subproblems, and solve them based on a correct and effective learning procedure. This digital neural network using reinforced learning could be the future of advanced machine learning, or AI could be considered the greatest artificial blueprint of the brain.

Although today the human brain is considered the greatest blueprint we possess, the possibility that AI will become smarter is high. Artificial upper intelligence could be upgraded with a set of neural networks working together until AI develops self-learning to improve and upgrade itself. This could happen very soon. Then deceitful acts could be detected by using AI. We could see the replacement of TSA by self-sufficient AI. A new system of screening could replace what we have today. A perfect, state-of-the-art graph of our bodies could be done with AI.

With a new system, when passing through, we would simply be scanned by a robot with infrared cameras that detect everything within the human body. No time wasted. Its accuracy would be perfect. In the future, we need not worry about superintelligent AI. What we need to be concerned with is how fast AI learns and what it does with it once it operates on its own. I can imagine a future when AI works overnight to solve our problems while we sleep. This could be the bright future of having an AI system working for us. The productivity, the progress, and the financial benefits could be our greatest human improvement. No more being scrutinized or touched by human hands in violation of our right to privacy under the pretenses of security. If technology is to be used for our protection, this is one way we could prevent and enhance our sense of human dignity and protection.

We still have much to learn about the insights of how a combination of the mind and the brain will help us improve our lives and create a better world for us. The mind helps us experience internally and externally everything existing in us with the help of the universe. Using the mind for constructive purposes expands our ability to connect to that field of energy in the universe. God only knows what we are tapping into by using a machine or AI to learn about our conscious mind. A strong AI can do anything better than humans, and this will give us a clue as to how the brain works. With a superintelligence at our disposal, we might discover that cognitive expansion is just the beginning. The odds that a machine can learn with precision how to organize data and structure it to make complete sense on its own is rather miraculous.

Any system programmed with a good algorithm can be learned by a machine to help us solve problems and complex situations. Intelligence is but random learning, and machines are very good at doing this. This creates the enigma as to whether AI can control and manipulating any system once its intelligence surpasses ours. AI could possibly convert and control all systems of information or programming on its own. It can reprogram or create its own algorithms and input and output of information by following a specific pattern of its own. If this happens, it will cause a threat to humanity; then we will have to question our sanity. All this can be prevented if we control all cognitive learning and, thus, process only what we deem necessary to our benefit. We must filter all the information we input for the cognitive intellect of AI to learn. However, if there is a well-designed interconnectivity between our brain and AI, the chances of maintaining control are minimal. Machine assimilation of data seems to be very fast. We may find that controlling a machine after it has learned to operate on its own is nearly impossible. This can happen because a machine filters information differently than we can with our mind, and it can learn much faster through repetition.

What measures would we take to stop machines from thinking rationally, making decisions for us, or reading our thoughts? We could easily fall under their spell and have them tell us how to love, what to do, and when to do it. My desires are for AI to make a more comfortable world for all of us. We all know this is not always possible, though. The ambition for AI enthusiasts is to create the ultimate and most competitive system possible to make a global impact, thus using it to benefit creators and users alike. This is one of the reasons why AI has become the most competitive of all the futuristic intelligence programming.

Algorithms and neural network programming have the potential to expand our minds, our world, and our lives and make us either rich or control our world. The question is, who is going to create the best algorithm of the future? How will it impact the world, and what benefits will the system create for us? The future may be uncertain for those who are not in tune with AI progress. However, the process is in progress and continues to improve continuously.

There are private as well as public AI communities working on the future progress of AI. Not all data is available to the public. How far we push the limits of the brain and the mind is only a matter of trial and error, and complications are the evidence of our success. It is my sincere hope that those in the field of AI advancement take into consideration the priority of man above machine. As new developments and discoveries continue to evolve, the need to become more creative will expand with AI. When AI interfaces with the human brain, our ability to learn and processing information will grow exponentially.

We are constantly searching for ways that machines can solve our problems. As a result, human beings are evolving faster than expected. The evolution of man since the introduction of computers has been faster than anticipated. Our desire for more efficiency and the demand from corporations, governments, and individuals create the opportunity for us to become more creative and intelligent. Ye, there those who believe that perhaps we've found the magic genie or that universal knowledge is open to all mankind. Undoubtedly, we have only just begun to use our mind and learn how to use creative intelligence to expand our own abilities with machines. Whatever this mystical learning, constant seeking, or dissatisfaction with progress is, I do not know. But there is one thing I do know for certain, and that is that we are never satisfied with who we are, nor will we ever be. We are consistently seeking more of everything to improve society and ourselves.

With the discoveries of Einstein, Edison, Tesla, Niels Bohr, and all those who have made an impact in our lives with their inventions and advancements, we have begun to understand that man has no limitations to create or reinvent his future. In time, man has realized that with his mind, he can create a better world for himself and society at large. With his new way of thinking, man has taken a leap of faith into his own destiny to make this a better world and expand the mind where no other society has ever gone before. However, there is one aspect of our growth and advancement that has taken a priority in our competitive world: the expectation for a more advanced tomorrow, a curious world with greater, faster processing capabilities than we have today.

Scientists, engineers, programmers, and mathematicians are in a race to invent a faster and more creative AI to incorporate with a brain. The quest for the ultimate challenges has begun. I am in this race to find out more about the workings of the brain. The challenge is to find a way for a computer calculation to helps us process data faster than light speed. Then the worry arises as to who controls all the information. How do we keep all the data safe? Who is going to be in control of how this information is exchanged from one system to another without losing control of it?

When the time comes, we will have to worry about competition, who will benefit from it, and how AI is going to be used for everyone's benefit. Will it be us, individual enterprises, the government, or corporate America? We want to know what makes us who we are and what gives us our ability to think, create, and be the best we can to thrive. We are now learning how AI is processing information from the field of energy and information. The more we teach a machine to use data, the more we understand how processing works. As we speed up the process in hopes that the machine will be responsive, we learn how far we can make this machine respond to our demands. Many of the trials now in progress have shown the capacity for machine learning and processing information, which give us back exactly what we are seeking.

These machines are capable of processing more data faster on their own.

What we have failed to understand is that machine learning is no different than our own. Our brain is a computer simulation. The brain can learn and process information just as much as a computer does. Undoubtedly, using a simulation with a computer-brain interface could teach us much about how our brain's neural network transfers and processes information. The future of interlinking is promising. However, there are dangers in programming a machine to act or think like us. The potential for any unexpected outcome is there. Eventually, once we have made more process or learn how to connect the two well, we may learn more about how well the two work together. There is big anticipation to see the outcome.

It is my firm belief that we are about to discover what machines can do under our command and even outside our control. I think of this era of AI creation as machines whose capacity to operate came from man and inherit the same habits and learning of mankind. Processing information for a machine is nothing more than teaching it to follow up with instructions. Then the machine repeats the process until it gets it! The only concern we should have about is, for the machine to learn from the process and to become smarter than us? This is where the field of information processing will be tested. If information is in the field of energy, this machine we program is also accessing the field to interpret and learn about everything around it with its own energy incorporated in it. We will have to question whether processing information is part of the field of energy or data input.

Studies have shown this is not the case. AI has the power to process more information than we have programmed it to. If there is one thing, we have learned in the field of machine processing data, it is that nothing is limited, not even machines. With this understanding of energy and the field of information processing, we can be certain that what we are looking for is possible to find. My only concern is how we are going to use it and whether we might discover more than we bargained for. What we are seeking is out there! How efficient our programming and intuitive thinking are is the key to finding it!

The purpose for creating AI to alleviate the present stress from our jobs is to make our lives free from corporate demands. We hope that AI will do for us what we are not capable of doing on our own. Not only could AI process information or work for 24 hours while we rest and live a more productive life because of it, but it can alleviate our present world demands. I see the future of machine and human interaction as one where all we do is set up a program to run all night or all day with all data to follow, as we have programmed it to do for us. What this means is that all the routine jobs people do today will no longer be done by human beings, but by AI.

Will this affect the job market, or will it simply improve our living standards to where we no longer work so hard to make a living? What if AI became so productive in processing data that it did our work for us, allowing us to sit back and benefit from the results? It has been estimated that AI will do 30% of our jobs in the future. Major transitions are happening every day, and our lives will forever be changed by them. Consequently, our jobs, lives, and even our thinking capacity will have to adjust to all the changes coming. Reality as we know it will never be the same for us once we have created a machine to incorporate our thoughts with.

What used to be the norm for us will be replaced by a new normal. Can you envision a world where our need for food, medicine, and all necessities for living conditions will be provided by a system or machine in the form of a simple liquid we take every day for sustainability? This system will benefit everyone, no matter their status or condition. My vision allows me to see and imagine a world unlike the one we live in today, a new world where harsh labor is done by machines while we humans enjoy its fruits. Why not? If we are presently inventing the future, we may as well enjoy its benefits. We create all the changes, inventions, traveling into space-time, and so much more, so why not take care of our basic needs?

In the future, there will be no totalitarian regimes, no more governmental control, and no more elections. People set their own rules and abide by them. If you don't follow the rules, you are out of the system on your own. For a society to change and transform, a new structure must replace the old. If we continue to live under the same system, we will have the same outcomes. To paraphrase Einstein, we can't solve the problem with the same solutions we used before. After all, we are creating a completely new world for the next generation. The inventors and creators of today are the best reflection of our future. For any society to advance, it should become better than it was before or else it will perish.

What is the purpose of searching for more intelligence, for the limitless, the unknown, or the need to excel? Why is humanity's ultimate quest to discover the unimaginable? The human spirit is never satisfied with the now! There is a constant search for tomorrow's new outcome and the eager need to achieve what some might consider impossible. Will we ever reach satisfaction, or will we simply create the blueprint of history with our discoveries in the hope of being remembered as the best generation ever to exists?

Unlike the rest of the animal kingdom, man is not simply satiated with his common life. Humans seek to better themselves, to excel, to prove beyond any doubt that they are better than their ancestors. Evolution challenges us to be better or more competitive than previous civilizations. One generation wants to prove that it is better than the last. Competition has always been a way for any society to supersede another. This is how advancement has occurred for centuries. Man has evolved, ascending from caveman to the most intelligent species to ever exist. In ancient times, they competed in the sports arena. Today nothing has changed. Competing to be the best among the best is still man's primordial need. Competition is the cause for all the transitions in religion, laws, governmental control, status, and even in finance or hierarchy.

I predict that AI could evolve into a form of control and competition for men of wealth and financial status. If this isn't the case, why would they invest in a project that they consider to be dangerous and evil in nature? I sincerely believe that this control over AI is not about creating eminent sanger for society but rather, control over who is doing what to benefit from it. AI promises great financial benefit for the most creative inventors of the future. Some refer to AI as demonic, yet they have a great interest in its development, so much so that they are focusing on what or who is making a difference in any AI programming or design. Nothing is further from the truth. The future of AI can be promising because it will give manpower over the majority, and thus, it will benefit those with an upper hand. In other words, if a creative genius develops an AI program that can impact society globally, we could become a financially stable society. What AI needs is the next Bill Gates, a person who can create a system or program that enhances society and benefits all at the same time but with less personal gain than today.

There is a great potential for future control of every aspect of our living with the use of AI. Imagine it as the biggest transformation ever to have taken place in the existence of humanity. AI will play a big impact on how our future could be managed by the powerful and affluent. There is nothing evil or demonic about a machine we programmed and control. After all, if we are creating our future with machines and a system of algorithm and data processing, the responsibility lies with us. We are responsible for what we build, do, and create. Nobody else is in control of it; we are! The key factor is how we, the creators, will process and use the information. How will this information affect us, our future, and the way we evolve while processing data at the speed of light? Can we simply cope with data processing at light speed? Or are we going to create a program that we can't control?

The future of data processing is big, and the demand will grow as we create more programs with AI. Likewise, these machines will eventually learn how to do their own processing from one unit to the other. Once these machines have acquired a high cognitive intellectual ability, perhaps they can do their own programming. Information today is available and fast, but AI can improve the processing of data and get better with time. If information is what we are seeking, information is what we will get. While processing data, the system might teach us one or two things. It can teach us how to interpret data from a system of algorithms incorporated with the machine as it learns to communicate with us back and forth. The machine will also have to fully process all the data we input daily, and just as easily give us an answer. If we can achieve this, we will have managed to make the machine think for us just as soon as we input data into a system.

If machines of the future can process data as quickly as we demand them to, they could possibly contact planets in our solar system through a data communication center placed on Earth or in space. Once the machine can allocate the satellite or communication center, it will relay back to Earth every minute detail it collects while communication is being established from afar. We will not need an EELT to know what is going on. Then AI could help us detect intelligent life on another planet by interpreting sounds or detecting energy.

Such possibilities are uncertain for now, but the future of AI has no limitations. Any electronic device with robotic ability can be used to help us search for answers to all the mysteries of the universe. If we discover another planet, could we find intelligent life or evidence of its existence in the past? What kind of discoveries would we find, artifacts, messages encoded on stones, or simply remnants of the last species to have existed there? The only inconvenience we might face would be to compete with other nations to make further discoveries and become part of history.

If this happened on Mars, the quest for colonizing the planet would become a race. By then, robots and machines might have acquired superhuman intelligence, and we will be directed by them. Things could get complicated, and we will completely depend on AI to make any future discovery on Mars or any other planets. In this case, all the data we gather from space will mostly depend on an AI system in space. Presently, we are not sure about what is happening on Mars. NASA highly secretive agenda is not for the public to know. Nor do we know what invested interest other participants in their projects have.

AI AND OUR FUTURE IN 2020

CAN WE USE AI to stop hackers from entering any private, financial, or government sectors? If so, how? Rational computation depends on the relative importance of computing and programming data, so it deals exclusively with the objectives of the command input into the main processor or bank. Once the calculation has been computed, recorded, and programmed to its precise value, in the future, if any other non-computed values are in question, the program will simply reject them. This means the programable data will recognize the intruding codes, where they originated from, and how they penetrated or filtered into the system. There is also a possibility that AI will give us the time and location of intruding hackers. If anything is wrong with the evaluation, there will be a request for a change in the program. This will alert the machine or computer.

It will be impossible to decode an entire program even if something is wrong with the coding. Any incorrect data will simply be rejected. Then a new program will override the old one. The main brain has a built-in system of computation that processes all incoming information before it is accepted and evaluates it completely before it is installed. With the main brain of an AI, we can apply its rational calculations from all inputs and process the data to its entirety. This could be done with the splitting of neurons and brain receptors in the AI brain to make sense of the computation while they detect any errors due to unorganized computation by the programmer.

This assimilation of processing and computation is done by a general server, which then detects any incoming data not recognized by the program. If all other attempts to save the data fails, any number of methods could be used to decipher the information. We can either save it or block the codes that have already been used. This would stop anyone from penetrating the main source of information. It is important to try and run the program and ensure the system has been clear. Once we have accomplished this, creating a new program with new codes and keywords could ensure safety in the data.

If any disruption happens because of hacking, there is a way to correct the overlapping of data coming in from other sources. First, shut down the system or simply divert the data to follow up on a new pathway to the perpetrator. How can we do this? To divert any system, one must first ensure there is a violation of the input codes. Then create an alternate site to run the data with a different code and watch how the system runs. Running a different system with a new code will certainly let the programmer know when and how the hackers are trying to enter the program and from where the hacking is originating from. Using a different code with the same system will secure protection over the original program while it is being changed or transitioned into a completely new program with different codes and language.

Once the new program has been completely transformed, it will be impossible for hackers to attempt to access its data. They will be locked out and will have difficulties getting out of the system. It will be like getting caught in a net when fishing. Disruption of any system is only caused by interference from other sources trying to penetrate the program. However, this could also be the result of the overlapping of data input. This means the system is taking longer than normal to process all the information that has been programmed into it. Even though we can change the coding or create a new program, the most reasonable action could be to save all data. In the event we are unable to do so, it is possible that all the data could be lost. However, a smart programmer always has a backup!

With the information at hand, one can design a completely new program with similar data from the previous program. Here, it is imperative that the programmer uses creativity to prevent any perpetrator from deciphering the new program. When creating any program or putting new data into a system, deceiving intruders is a must! Thinking ahead of the hackers is the main point. Deception, deception, deception!

Cognitive intelligence is necessary to protect the programming of any AI or machine. Yes, both! The machine must be equipped to process any and all information input into its system safely and securely without being overpowered or outsmarted by intruders or hackers. Providing adequate data is imperative for running a secure program as designed. All data must have a reliable source. Any data received has value and intelligence in it. If we want confirmation of any incoming data, first, we must run it in a backup system in case it is a virus or has malice intent. Then we must evaluate its content. When dealing with information from anonymous sources, caution is essential. Once we know where the data came from, we can determine its purpose and motive for entering the system. Nothing can be left to the imagination when dealing with unknown incoming data.

There will be an attempt to incorporate other systems with the original, but if the program has been designed to detect all unrecognizable data, it will immediately alert the system that something outside the norm is happening. It could even shut down the program as a precautionary method of dealing with security violations. This does not mean the system cannot be reinstated. It simply means that it will need a new code. No one except for the programmer has the authority to do this. Only one person is assigned to consistently deal with the security of the program. Creating a program or AI of this magnitude means that we have used intelligence with cognitive programming in it to be smarter than us or to prevent any possible corruption from any outside programs.

In our search to discover more about ourselves and the ultimate cognitive intelligence of our mind, we have not only discovered a better future for ourselves, but we are accelerating the phase of our thinking with our ability to tap into intelligence through a magnifying glass of the unknown, the invisible power of intelligence in our curious mind. AI is men's intellectual creation as part of our own imagination, invented by us from a field of unknown intelligence available to all but only understood by a few.

Although, we are all creative individuals, not everyone is interested in contributing to ideas of science and technology to enhances our future. I personally know for fact that ideas are a great commodity for hackers and business entities alike. Today, there are those who steal information from naïve people like me who have less understanding of technological advancement but have a creative mind. They consistently scrutinize social media and perhaps personal emails to get ideas. This week of October 20th until today the 27 of October, I have written my idea about vision three times in this manuscript. Then, for no reason, it completely disappears from my writing. Also, the idea about subagent is been removed several times. Persistent is what I am, I will not give up. Whether I am right or wrong, if my ideas seem to agitate anyone, there is a big reason for it.

Perhaps, these ideas have great value, or they are been used by unethical people who steal information. During the process of writing about this AI pros and cons of the future; I have had several computer breaks, my computer going blank and emails disappeared. It is been said that if you have to break a wall to get to the other side, but you can't; perhaps you should give up because there may be inconveniences on the other side. However, persistence is the only thing I know that pays off in the end. Giving up is not an option for me. I believe I heard this one from Mr. Elon Musk. I have dedicated almost three years of hard work to come up with all the data and possible outcome of incorporating AI with the brain, or how AI can improve the future of us. Having imagination does not required a degree. Becoming of age does not mean your brain is dying. Being creative is a gift, and I will not let my legacy die because of the impertinence of others.

CAN WE MICROENGINEER A HUMAN BRAIN TO THINK LIKE AI?

MICROENGINEERING A HUMAN brain to think like a machine will forever change the way we live. How can we make the human brain perform at a higher level than robots or machines? To micro engineer a brain to think, calculate, and have human intellect as well as machines, such a brain must act at a superintelligent level. To think faster than any human, this brain must run at full capacity, faster than humans see things, at the speed of light or greater. It will have to receive information from a source that is greater or smarter than itself. Such is the case with a computer or any program we put together.

This has nothing to do with cognitive consciousness nor intuition. Rather, it has to do with our intelligence being put into a system that then transfers such data to a human brain and does for us exactly what we ask of it to do. Neither a machine nor a brain can think without any input. We need books and knowledge from others to learn, and so, too, does a humanoid machine brain acting upon our demands or programming. Every program we create begins with a thought, one we create and then put together.

When it comes to microengineering a human brain, we are

creating a connection with a thinking brain and not simply a reasoning machine: a programmable system repeating data and mimicking interaction with us. To microengineer a human brain to think with a machine, we must first understand the deepest basic reaction between the neural transmission of data from a brain to a machine and back. It is also primordially essential to understand how the machine will process any data when it has received it. We hope that the process of transmission of data will go as programmed; however, until several trials have been conducted, we won't know the end results. The key point is to understand how the brain processes information and then program such information into the machine and hope for the best.

Wouldn't it be beautiful to have multiple results from the machine at once? Such input will create greater hope for the future of brain-machine interfacing. Could we create a brain that can respond to us with accuracy? If we can make a new set of neurons that can communicate with one another—in other words, if we can create a new brain—what is the possibility that we can train it to think like a machine and have similar results? This is just a thought into the future.

This, of course, would be the reverse of AI. My thought is that it depends on the purpose behind it all… If we attempt to create a humanoid or human-machine brain interface, we must think of the opposite to understand the adversity of its creation. We cannot possibly understand one without creating its opposite. It would be like electrons communicating regardless of the distance between them.

Nevertheless, there is a possibility that the neurons of both machine and brain could recognize and interpret music, images, and even what is real or fake at the same time and without any errors. Thus, this will create the first impression of communication at a distance between machine and the brain. This could not create an adversarial neural network, because both have similar data at the same time. There will be no competition between them. Anything is possible. At this point, we don't know the limitations of the neural network's deep learning process.

Because AI is capable of learning more than we expect, in this early stage of programming and data processing, we expect that it will be able to create its own language, conversation, logic, and common understanding with precise detail—or develop codes of its own that we humans do not understand. How its main operating system works from hardware to hardware is still unknown. Supposedly, at this stage, AI will learn from our own deep emotional intelligence. If AI is reading our body and mind vibrations to learn more from us and about our human behaviors, at this point in our evolution with AI, we are learning how to transcend our biological needs. If such transitions are taking place between the machine and our brain interface, the progress of brain-machine intuitive augmentation between the two has begun.

This cognitive algorithm between the two can simply mean that we have succeeded in obtaining effective thinking, acting, and communication between brain and machine. Unlike a computer that analyzes and computes according to the questions asked, this brain-machine interface simply responds mutually as one thinks before the other or vice versa. This achievement means we have taken a step forward in creating a connective pathway between the brain and a machine with great success. The pattern of recognition that allows both to have intuition, strategic thinking, and logic has been established. The system of microengineering the two has created a new way of communication between a human brain and a machine.

The doubts many have created about AI and what is

possible can only impede our desires to be successful at any task we endeavor. How do we make these two interfaces smarter and create a superintelligence? The brain has a thousand billion neurons. That's a lot of brainpower to do anything or interact with another superpower, such as a machine. Working on memories and learning is one way of getting smarter. The more we teach a machine, the smarter it will get. Because a machine doesn't need to rest or relax, it is capable of learning much faster than us, and it can teach us what it has learned in a shorter time. While you sleep, a machine can be active and learning. When you wake up, the machine can give you a brief explanation of what it has learned in 20 or 30 hours.

If you are wondering how your brain can retain this much information, don't forget that it is interfacing with a machine. Because your frontal cortex is involved with learning and memory, it will activate twice as much when it is necessary for you to learn new data with the machine. Some areas of your brain will be activated only when necessary. It will be like learning to use your brain anew. This is what interfacing means: a new programmable system of integrated brain-machine interaction for thinking mutually at the same time with similar data available for both to do cognitive processing simultaneously.

Would the machine be faster than us, or would we have to increase the speed at which we process information? Can one possibly imagine what it would be like to function as a machine? How would one ever shut down the brain from overthinking, and at what level would it generate information? These questions are essential for us to understand thoroughly how our brain would react and, thus, generate electrical impulses to transmit data at the speed of a machine or AI. When dealing with a brain-machine interface, the outcome can only be predicted if we consider all possible implications.

Hence, our experiment is essentially a new one; there would be complexity as well as a constant adjustment from both. The mere idea that we can get all the wires and our neurons to function at once is rather exciting, and not knowing beforehand what could happen adds to the anticipation.

Whether we use a machine or a humanoid brain, both would need to be programmed or set within our commands. A system can only run itself if it is programmed with all the data to do so. A superintelligence does not create itself; it is built by humans. But it could also learn on its own and create its own system. At a certain level of intelligence, it, too, can program on its own. We underestimate the fact that learning is a process from beginning to end until it is fully accomplished. When a new program is designed, we don't know the results until it is fully executed. The same applies to AI learning. With AI programs, data and programs are input together and then later implemented to get a possible favorable result.

We can build a super-brain by joining it with a machine, and we can operate it from the central nervous system, or we can enhance its intelligence with data input. If we understand the consequences of this, we can manage to keep under our control. In this case, the brain and machine can communicate together, the machine can calculate, the brain can reason, and we can put as much information into it as we want. There is only one problem. We would be fully responsible for any issue arising from our poor judgment, programming, or planning. A brain of this magnitude could easily begin to think on its own. It is possible that once the program has repeatedly been running, further intelligence could be enhanced by either the human or the machine. This means that one or the other will develop more intelligence and think and act on its own.

Current robots and AI machines can imitate humans, have a sense of humor, and beat us at games, but we don't expect AI to be able to learn to read our deepest emotions. Once the interface has taken place, who is to say that AI is not going to read our behavior, emotions, or even our sexual desires? Then what? Could it be that AI will mimic our feelings when we are sick, sad, angry, or despairing? Can interfacing create a mutual brain that not only communicates like one, but also thinks as one? This can be called a unified set of intelligent and cognitive data because the machine has learned how to program itself from our rules and/or data. We can assume theoretically that computers of today do just that with information or programing in them. If our intentions are to create a new program, we can unify to operate as one. We must ensure that all programing or algorithm is simultaneously corresponding with the other to get a perfectly organized solo system. Such possibilities must be considered when dealing with a new system or programmable brain interface. We have no idea what to expect until the program has been completely designed and finalized. The odds depend on what we want to achieve with the program, or how do we want to use it? We can create one or two simultaneous programs. Each one design for different purpose. One to relate data, the other to maintain mutual communication between two programs. If interfacing purpose is to achieve communication only, multiple programs can be created for a variety of reason. This will secure success of one program over the other. If one program does not respond as programmed, there will always be a backup.

WILL THE HUMANOID BRAIN BE EXTINGUISHED IN THE FUTURE?

ALTHOUGH OUR DREAMS and expectation of AI are high, the process to override the brain, nerves, as well as the muscle and brain neurons would exceed its capacity in a short period. What do we do with a humanoid whose time has expired? Do we kill it, let it become one of us? Do we use the remaining data from the microchip in the brain, or do we simply reduce its mental acceleration by putting it to work in a different area? The answers aren't completely clear. But nothing says that once we have created an interface, new interfaces will not be formatted elsewhere. Even if we extinguish them, will the rest do the same, or will they simply continue to experiment on whatever else they can to conquer a higher goal? Can we use them later if needed? I suppose if we have a critical situation in our hands and need extra help, they can be of use. We can always go back and use the humanoids for problem-solving.

The future is unpredictable. The remaining humanoids would be overworked, mentally stressed, physically exhausted, and most likely mentally overused. The human brain can only take so much information processing. Although it might be connected to a machine, it will still function with the rest of the body. After it has delivered to us all the data, we have input, we must find a way to deal with it. How do we slow it down? Are we going to deprogram the brain to slow it down? What would happen to the brain once we decide to change it back to its original status? Would it slow down neural processing, or would it simply adjust back to its normal state? Brain enhancement is possible; however, when use as a machine, the regeneration of its neurons and/or processing information will slow down. Studying the brain after trauma or damage can teach us how to deal with this problem.

Once we have conjoined the brain with a machine together, desensitizing one from the other may have some consequences. When dealing with a human brain and machine interface, the transformation, as well as detachment of either, has positive or negative results. Doing the trials now would give us an idea as to what to expect later. Today we are at the stages of trials with AI. Using all important measures will prevent errors and, thus, create certainty with what to do or not to. As AI is integrated into our lives in different areas, the more we will depend on them on a day-to-day basis. However, the more we know about how to deal with any inconveniences AI may present to us, the better we will learn to coexist with these machines. After all, our future will entirely depend on the functions of AI. We are already seeing its use in many areas of our lives, and I need not mention any further.

A machine takes longer to coordinate than a humanoid. A strong superhuman performs better than all humans put together. Why? Because when humans work together, they create complexity, while a single superhuman will not. Of course, we are referring to a designed humanoid, one we can control and manipulate at a distance. Today a computer can beat a human in chess. Non-human machines can outperform humans, but a humanoid is easier to control than a machine. There is no threat.

Safety, though, is essential when dealing with security or privacy issues. Today we have hackers threatening our security and privacy with machines whose intelligence can bypass any code or program we use to stop them from breaking in. This is a good example of how machines are being for harm. There is no doubt this is possible. However, I am a firm believer that we can put an end to this. The only reason we get hacked is that we tend to use one password without changing it for a long period. Of course, they will hack it. There must be a way to be smarter and more cunning than hackers. If there is a way, we can avoid this, why not do it? It is not that they are smart and can penetrate any security. No, it's more likely that we do not protect ourselves enough to keep hackers away from using machines to take our system down.

Our safety and security are our responsibilities. Therefore, we should figure out how to take down our opponent the minute the machine gets hacked. There must be guidelines we can follow to lead us to where the link originated from. There are many ways in which we can protect our safety. We get little cooperation from the authorities in preventing such personal violations or atrocities. We should not be the ones fighting to protect ourselves. Our lives and security should be protected but not monitor constantly. Freedom is not losing the battle against your enemy and being defeated. Rather, it is securing your safety and security against all enemies.

HOW WILL OUR LIVES
IMPROVE WITH AI?

SCIENTISTS HAVE FOUND that multitasking and the use of cell phones and computers are making us depressed. This is where AI can come in handy. Our brains can only take so much of the daily ongoing technological engaging going on today, but if we can create an artificial intelligence to release the mental pressure, this would be one big step towards progress. It's a fact that our emotional intelligence has been affected by our addiction to social media and computers. Nevertheless, some criticize me and many others for thinking outside the box. AI enthusiasts are passionate about learning and writing about artificial intelligence. I am sure they will be the first to benefit from all the ideas about AI. When I talk about building a humanoid, I am thinking about improving our ability to communicate and explore avenues of unlimited possibilities between man, evolution, space-time, and the challenging questions we have about the universe.

When it comes to what is possible with cognitive flexibility, what we can do with our mind, our brain, and artificial intelligence? Imagine, for example, that we use multitasking as a way to improve the human brain, solve problems, and create a new way of living. We could process information and solve problems faster. We could heal patients instantly. Or we could use problem-solving strategies to live happier lives. In the process, we must improve as well. We cannot possibly create a better world in which we, the main recipients, are left behind.

If instead of worrying or using the subconscious mind to deal with our daily problems, we used it to solve problems or find solutions to different tasks or new inventions, we would become better thinkers, more functional, and maybe even much more creative individuals. Cognitive flexibility serves us well when we are uncertain about a decision, a problem, or a solution for a better outcome in our lives. If we want to become better human beings, we should learn to let go of the constant reusable software of the old subconscious mind and become more creative at producing better results or solutions to problems. Picture yourself in the future using your mind not to record old uncertainty or wrong decisions but rather using it to find a new way of using your conscious mind to solve your life uncertainties.

In my personal observation as to how things are progressing with AI, I think that cognitive flexibility will help us transcend from the subconscious mind to a new way of thinking in a more productive and beneficial way. In other words, we would become better thinkers by finding ways in which we could teach, learn, and solve problems in a constructive manner, thus switching our minds from the old way of thinking to a new way of solving solutions.

Although cognitive flexibility has made us more of a multitasking society than ever before, its benefits outweigh its complexity. If we use our minds to create a better world rather than worry, we will bypass the acceleration of time demanded with the expansion of space-time with dark energy. Cognitive thinking and flexibility help us learn how to use the mind in a multi-faceted way, to think better or use the mind in ways that are more productive than not. Our efficiency in learning would accelerate, improve, and help us think outside the box.

First, though, we need to incorporate consciousness as our essential aspect for being. Only then can we understand the correlation between being human and connecting to that which makes us greater than just human beings. This is our altered state of being, the complete connection to all in this universe that makes us part of the invisible energy all around us. Then, together, our minds will process information faster because our energy will be amplified, helping us expand the way we think today.

Cognitive flexibility does not require a high level of psychological education; rather, it requires that one think irrationally, or outside the box, unlike all others. To help one think outside the norm, all you need is to learn how to solve a problem differently. For a machine to learn to do this, one must think of several ways in which a problem can be prevented or solved by a humanoid or machine brain. When all the solutions are coordinated, the problem is more solvable. This is how we, too, can learn to solve problems coherently. Like a machine, our brain should learn to move with the essence of time.

If we only focus on the idea of transferring information from the brain to a machine, and vice versa, we are limiting the brain's capacity. The idea is to enhance brain activity, improve our lives, and create a better world to live in. This is how we do this: we should transfer information from a human brain to a machine with a microchip or other device. Then we can say we have conquered everything! This will be a phenomenon in communication. Our communication today is our primary source of survival. Whether we do it via cell phones, computers, or any other device, this is the present world we are living in today. Why not expand it with AI to increase our status as human beings and navigate the universe with ease?

There are no complications between the brain and bits of information from a machine. We are already using this today. The process of thinking from machine to a human brain already exists. Stephen Hawking is a great example. He was connected to a computer that translated for him, as he was unable to speak. How they linked together was also part of AI creation.

Quantum physics says we are part of the elements of the universe, such as atoms and photons; we, too, are incorporated with the field of everything that exists in the universe. Our thoughts work at a quantum level of the field of energy. In the quantum field of energy, the brain can increase its capacity to transmit and receive information. How this happens is up to us to decide. If our consciousness affects the behavior of subatomic particles, so can our thought process affect how the brain is interacting with the field of everything in the universe. Every thought intertwined with the field creates an amplitude with other fields around it.

Does that mean that AI thoughts can improve with the field as neurons multiply and interact together? There is a great possibility that as AI intellect improves with time, it, too, may interact with the field of energy and create more from the source. Are we debating whether AI can connect to God? If the source of everything that is was created by God, then it, too, connects to all other creation. If AI is to have feelings and emotions like us, it, too, will have the capacity to recognize a source greater than itself.

The question remains: will AI understand what it is? The answer is clearly no. A humanoid cannot have a concise understanding as to what a subliminal being is. It only can understand that which it is exposed to by knowledge, learning, and training. All else is but speculation to it. Even a rational person with no knowledge could not understand the interpretation or meaning of what God is. Since we have not decided at this point whether AI is capable of loving or having sex, we also must exclude it from having any knowledge of God.

WHO CONTROLS THE FUTURE PROGRESS OF AI?

I UNDERSTAND THE concerns of those who worry about the outcome of AI. What they should control, though, is not the communication level but rather who or what has the authority to do what. We cannot anticipate whether every invention will be a challenge or not. What is important is to know the purpose for which this new creation has been adopted; is there a greater purpose for the invention that would help humanity in any way? If there is one thing I know about the AI, is that it will not exceed human intellectual capacity unless we design it to do so. We hold the key to its progress, and we can control all of its functions, human or not. AI is bound to learn on its own. Because we are the ones creating the improvements and supplying the brain, we have the key to the master plan. However, nothing is perfect, and we are bound to make some mistakes along the way. These mistakes will teach us how to improve AI and what not to do.

If the government intends to keep AI a private matter, this will only create more curiosity. This is the nature of human beings. If we develop a program within the system of algorithms that can generate more information from the body, we have managed to control with precision what the outcome of all answers may be. On the other hand, if the system is designed to give us only precise calculation when we need it, we stand a chance of corrupting the system by overriding the brain. In other words, there could be an overlap of brain activity from the overgeneration of information from the mind to the brain, exhausting the system.

Artificial intelligence has numerous potential benefits. One of them is that we can use it to detect weapons in schools, direct traffic at the airport, and scan air travelers instead of using X-ray machines. AI can also detect anything at 360 degrees. We could use it to inspect airplanes before departure from the gates to ensure there are no bombs. AI could become the primary security device for all areas that require safety. If the system is ever corrupted, the only thing we can do to stop any further intervention from hackers would be to shut the system down and then remove the microchip. The second choice would be for AI itself to corrupt the system by using a diverted code to stop anyone from hacking information. AI is smart enough to do this and more.

We will train AI to act in case of a security threat. Now, imagine a humanoid machine making drastic or life-or-death decisions for us. Can we? The future of big data or AI is information turned into knowledge. Therefore, we call it a program. We program a system, brain, or machine with sensory input. This information is immediately available based on the running of the algorithm. As we pile up information into a microchip and then into the brain of a machine, we have accumulated a massive amount of data that can help us improve our programming. The more computational tools we have, the more we can observe the different patterns as we create a brain with this much data.

With the first observation after inputting the data into the brain, we can see how much information the brain retains and can translate into reality. How much data can we input into a brain-machine? We don't know to what extent the brain can hold data without being affected. The main network in the brain will help tremendously in empowering the brain to keep information active as we continue to program a better system of connectivity.

Data in our brain is like a treasure we discover. The more information there is, the more power the AI requires to improve, create, and penetrate highly secure sectors with its program. Our planet is like a nervous system creating new nerves and growing as our cells with information and data. The data volume is becoming vaster than all the grains of sand in the world. The more data we get, the more difficult it is to solve big data. The function of AI is to help us create a new system to change and compete with our present demands. This is where I see a need to find what we have been looking for. The invisible energy would be seen as the "big energy," BE, and "after energy, "AF." This means that things will have to be quantifiable. But before we can get there, AI will help us solve some of the biggest problems we face today with computers and hacking. If AI can compute data, it can also learn to decode encrypted information.

If we think of AI as being part human and part machine, with superintelligence capacity, the benefits outweigh the cost. Because AI will show no emotional response in case of threat or security, it can bypass danger as a normal human being or infiltrate any perimeter without been detected. Imagine if we had an agent who could bypass any security by simply observing the area and finding a gap to penetrate without being noticed. How would this humanoid develop a sensory alertness or detection of danger no matter when or where it is? Would we need more than one to protect us? The answer is no. One single humanoid can do the job, especially if we are

counting on it to not be noticed in cased of danger or threat to our security.

Security with AI can be of nay nature; like cyber security and others. When all this is done, who will control what?

AI, 3D INTERACTION, AND NEW INVENTIONS

AS WE EVOLVE with this new energy and AI, our interest in network associations will improve, and so will our lifestyles. A 3D game will be invented where one will participate as an avatar or cartoon character and act, talk, move, and react without the other person knowing who you are. This will be played as a form of game, but the people interacting are human beings. What would make it intriguing is the nature of the game. You will have to qualify to join. Also, the participants are revealed only by those in the game. You can agree to play the game or refuse to participate.

Before you enter the game, you can request who your partner is. The game will have a sexual, technological, scientific, or even psychological nature. The participants are adults. They must be qualified to participate. An ID code will identify them as they also use their voice recognition to enter the game. If any of the players attempt to use a voice previously recorded, AI will recognize it immediately. The device trained to recognize voice patterns. During the game, a player may request your presence. At this point, you can accept for monetary purposes, or you can opt to go to the function or donate your funs to a charity. How cool is that!

Although it might look like a game, the players or characters are real. We can call it a human simulation or 3D games. There is no danger in the game; acceptance of invitation means that all pertaining details and information from both participants will be set aside until they have completed their meeting or date. This will only take place with trusted and reputable members. Qualifications will not be based on status, but rather on the ability of both players to maintain the secrecy of the game and personal details.

I think a 3D game is a fantastic way of passing the time and getting to meet a prospective future mate. I believe the future of dating is about to change and we will no longer require a person to be there to enjoy their company. Expressing our feelings will not be difficult. The person you choose knows who you are, and both of you must agree to play this game together. There is nothing unusual, unacceptable, or even kinky about it—except, of course, what the couple decides to do or communicates on their own time.

AI will help us solve problems. It can stop any war from happening by using tactics that help us deal with the enemy with prudence and intelligence. The alertness of weapons can also be useful. When it comes to future wars, I believe that AI will be the best choice for problem-solving. We can uncover the secrets of our enemy by using AI as our primary strategist. After all, we are talking about a humanoid with superintelligence. Developing an airplane that can fly at the speed of sound and a spacecraft that can penetrate the angles of the universe undetected will be easy to do.

Imagine an invisible airplane made with energy that can fly undetected. No one can see it. We only know that there is energy flowing up in space. This would not be noticeable by the public. Energy from what source? Perhaps, energy found and extracted from a black hole. Yes, a black hole is so dense that even light cannot escape. Anything that enters its core gets completely devoured. Because this energy is invisible, we can discover its gravitational attraction and use it to create energy in space to do the same for our own benefit. Of course, we would have to know how to pull matter from this energy content. I don't know if we will discover this within my lifetime, but I do know that somehow, we will find out what is inside of a black hole and duplicate its gravitational attraction to improve our travel in space. Invisible energy will be the way of the future. The mystery of its nature is what will make it useful. Whoever discovers this invisible energy first will have the upper hand in space travel and future discoveries. This is where the term "The sky is the limit" comes from.

AI will know what materials to use or create new materials from scratch. It will let us know what areas of the universe to avoid. How can AI do this? Well, think of it as having eyes that can detect at ranges we cannot today because the installed microchip would allow its senses to exceed its normal capacity. The idea that a machine-like brain with human intellect can do such things sounds incredible! If we manage to conquer this, we will have created the ultimate dream for man.

Questions remain as to who will benefit from the creation of AI and how much will it cost to put it to use? The answers are either those who can afford the cost of its use or the government, which might intervene and make it a business of its own. I believe that once we have tapped into the chemical reaction of the brain and the body as we increase its normal capacity, imagination will flourish like a burning sensation in our guts. Imagine this: if cells can multiply, our thoughts carry energy, and our neurons multiply by the millions, what else can we not create with our imagination?

It is known that reactions from the brain can happen in a matter of seconds. Consequently, how can we control a brain whose intellect is greater than ours? If there is a deficiency in the creation of data information, could AI rebel against our command? This is a chance we take by creating a machine-like brain that can react as a human and machine at the same time. We may need to increase muscle capacity to have a fully functioning body that can respond to the brain. When I think of the movie with Arnold Schwarzenegger, *The Terminator*, I think about AI. Since creating an AI that can react, think, calculate, and give us the correct translation will take ingenuity from us, this complex system will have to be able to do what we are not capable of doing.

Can you imagine AI solving poverty and how to provide medical assistance in distant locations in Africa, Asia, or the Middle East? Brain-computer integration has a better possibility to help us solve any kind of problems we may face. Doctors, lawyers, and other professionals will not be needed. There will be a solution for every illness, every pandemic, every medical adversity we face. All court cases and records will be accessible to those in the private sector for conducting cases in or out of court.

Can you imagine that one day we will not have to leave our houses to go to court? A network or virtual reality and 3D will allow any or all cases to be heard right from your home. Schools will no longer be necessary. Children will have the freedom of learning at home with the guidance of a supervising authority to control what they learn online. Their curriculum will be assigned online. Tutoring online will also be available for free. Yes, free.

And one day, nursing homes will no longer need extra help from nurse's aides. A robot or humanoid can do their job. AI programmers will design them to help with medication, in and outpatient care, or any other physical interaction. Medication will be set hours before the nurses leave, and the patient will receive his medication on time. If anything goes wrong, AI will send an alert to the head nurse, who will then either show up or recommend seeing a doctor. There will be a doctor in house 24/7 in case of an emergency. Surgery will be done in one wing of the hospital and recovery in another wing. No two tasks will be performed in the same unit or wing. If a critical situation occurs during the watch of any artificial intelligence, the matter will be handled according to its urgency. Of course, AI can handle this and much more. Everything will be handled according to its programming.

The future of artificial Intelligence looks better than we think. Any problem we are facing today can and will have a solution with AI. The purpose of artificial intelligence is not necessarily to act like a robot, but to be of help in any area in which we are currently facing difficulties. Because machines have a greater capability to calculate than human beings, we can incorporate the reasoning and intelligence of a human brain with a programmed machine, and together they can create miracles in our daily living. The results could be miraculous. Imagine not having to scan our bodies anymore at the airport; instead, AI will do this for us. One look at the person, and the entire body can be scanned, including the internal organs. No one would be able to sneak weapons or unwanted tools into the airport—and no more patting asses or touching in any inappropriate form.

The next advancement in technology will be a house designed to have all the connectivity we need to be connected to the outside world. The internet and telephone will come directly to us. A 3D system will keep watch on your entire house while you travel or are at work. The system will give you daily reports. The important thing to understand about artificial intelligence is not that it will translate at the speed of light; rather, it will help us complete more in less time.

Artificial intelligence will be very efficient. A machine-brain combination put together to excel at many tasks is like building a superintelligent humanoid. The brain cannot process at such speeds without the help of a program, a machine, or both. The central data system will be the head brain for the entire program. In the sector of offices and paper transactions, artificial intelligence will help us process more papers in a short period with precision because of the program already stored in the system. Insurance, health care, doctor's appointments, as well as management, reports, and even schedules would be done by AI overnight. Consider it done when you get to work and AI has already scheduled your day, your patients, and your meetings for you. All one will need is a complete report. If you need to make any changes or cancellations, AI will send an immediate message to the corresponding person. Expect accuracy, efficiency, and even thank-you notes from all your prospect clients. Nothing will be left to the imagination. All matters will be taken care of.

What kind of brain and computer can perform all these tasks without error? The truth is that a combination of both would exceed our wildest expectations. We can begin to call it, "The business of intelligence." In the business of intelligence, everything would be based on the performance of a humanoid machine like our brain. Who is going to outdo whom is the question. Business meetings will take place with your boss or CEO present in the form of a holographic image. If the person is out of town, there will be no need for computers; a program will be designed to have you communicate back and forth with this person as if they were present in the form of an image.

Not only will cars drive themselves, but airplanes will fly themselves with the help of a computerized system from the ground – or ground control. There will be no accidents; the airplanes will land safely as they will be guided by an autonomous computerized system. Of course, this will take time, but I see it coming in the near future.

WHAT WILL HAPPEN IF AI GETS SMARTER THAN US?

ARE MACHINES GOING to outsmart us, or are we creating our own destruction? What kind of work will we do when artificial intelligence has taken all our jobs and we are left with just thinking? The predictions are that we are going to have to work on ourselves if we want to survive what is transforming our future with AI and technology. Are we going to be the operators of machines, controlling their every move and commanding them to do as we need? Or are we simply going to sit back and watch AI work in our behalf? The truth is that the future of humanity is bright and, at the same time, bleak. Why bleak? If everything that we do today that requires thinking or performing a task will be taken over by an artificial brain, what does the future hold for us?

Scientists and programmers feel that creating a machine with the ability to translate languages, form concepts, or solve problems like humans do will be difficult. I, on the other hand, think that we have what it takes to incorporate a brain with a machine and make it do as we desire. Because scientists and researchers worry about the outcome of a humanoid programmed to communicate at a human level of thinking, we must find a way of connecting this to a human brain with nanotechnology. The brain is like a wired system; it can work with an electromagnetic vibrancy that increases neural activity. This is a way of improving thinking functionality in the neural network. It takes understanding of neuroscience, connective science, and psychology to understand how to put a brain and machine together. It is ingenious.

This is one way of improving the knowledge or awareness of the mind. We first generate energy, then we use the mind to think, and then thinking brings awareness into reality. The brain is just like a machine. If we use a system of repetition, it follows. If we use a sense of awareness, it becomes more and more aware of what is and is not present. The brain works like a mirror: it will mimic what we teach it. It will think according to what we program inside of it. And it will follow the routines we teach it.

Nevertheless, machine superintelligence will be visible everywhere, from machines building other machines to thinking and acting on behalf of our commands and sometimes not. For example, programs are being built today to do precisely what we have set them to do. However, there have been cases where the AI has taken the upper hand to act on its own by using thousands of hours to learn the program. If AI continues to learn at this pace, it can take over any program in time and bypass our intelligence by learning how to make sense of any input or output in any program setting and make any correction or solve any problem we may encounter. One good example would be the SpaceX shuttle launch and landing complications.

The possible outcomes of using AI for any future project, whether it be space exploration or simply reconstruction of human DNA, are inexplicable to us. As SpaceX attempts to send humans to Mars, I can't help but think of the future possibility that those who are older may want to go on a trip to Mars with the conviction that one day, in the future, they will return. Using AI to make an exact copy of the DNA of a person going to Mars and save it for the future reincarnation of that person would be like them being born again. Can we even fathom the idea that one day, we will be able to replicate ourselves into the future with our own DNA? Imagine AI screening our DNA to the minutest detail to make a copy of it and save it for the future. Such a thing is already possible. It would be like creating reincarnation in human form. If the old DNA is used to reincarnate the soul, this soul will experience the same memories as the past one. Or the soul might split into two alternate realities. If this is the case, we will discover that not only can we use our DNA for reincarnation, but memory and/or consciousness is stored in it. Otherwise, an alternate reality will only be experienced in the present, and not past events, memories, or consciousness. The past would be completely forgotten. Since we do not know to what extent DNA or consciousness can be manipulated, the potential could be limitless.

Difficult as it might seem, we can reproduce a replica of DNA, have a different type of consciousness outcome, earn a new soul, and have two different identities. This sounds creepy, but it could end up being the result of our DNA manipulation or recreation of a new human entity. What we are playing with is fire and magic at the same time. No one knows the outcome of such reproduction. What we do know is that we are trying to connect the dots of all the questions we have about humanity.

Using it for this purpose has not been thought of yet! Recognizing our potential can only help us create a world unlike any we ever thought to be possible because we can use the mind for a better purpose than for destruction and build a future we can only imagine with our mind. Can AI simply take instructions to do whatever we ask it to do? If so, will it follow through without failure? If we have managed to build a machine that does whatever we ask it to do, we can say that the future is ours. Given all the possibilities, we hold the key to our future. Nothing says we can't do what we think. Thinking will have to be taken to another state of being—in the sense that we will not only have to think about ourselves, but about all humanity in general. The world will encompass humanity in its entirety.

Will it be possible for AI to detect and correct any malfunctions on a space shuttle as it orbits in space by simply overwriting what is happening? Instant detection, instant correction, or overriding the problem. No time to waste. Wouldn't it be a great idea? Anything is possible if we can think of it as a solution to a problem. Even if we must use a different alternative to solve those problems, there is always a way to find a solution.

We are setting computers and machine system programs to accommodate us. If we, as human beings, can change the way we think, so can a machine or brain we program to do so. The question remains as to whether we can make this machine-brain humanoid as conscious, compassionate, and aware as a human being. Anything we do is possible for a machine to do. Repetition is the mother of probabilities. We are the creators of every machine operating today. We control the system, we design its purpose, and we do it all with our mind, our brain, and our thinking. We can teach a machine to think for us, feel, and think some more. This will create a form of pre-programmed behavior. Then all actions and memories will follow.

The brain is like a recorder: it can record thousands and thousands of bits of data and process the information. We all know neural impulses cannot process as fast as any machine does today. What we can do to improve this is unknown. If we can use impulses to help the brain perform like a computer with a neuron enhancer, this will accelerate the way the brain processes information, thus, increasing information processing. The only thing we will need is the mind and information. How we program this information is what matters. When the integrated brain, mind, and neurons work together, we can incorporate any data or program into a machine to have it produce a desirable outcome.

Before any program or algorithm, the brain will have the capacity to process all the information at the same speed and thus recognize it with precision. In such a case, the brain interfaced with the machine will have to become like a machine to operate efficiently. Given that the neurons can process information fast but not as quickly as a machine does, the machine or AI will have to guide the brain, or the brain will have to learn to read and process information as quickly as the machine does. To make this happens, it will take either enhancing the neurons or improving AI to think rationally like we do, or we can create and produce a thinking machine that requires only input and guidance. Then this machine can process as efficiently and quickly as possible to produce all answers in a fraction of the time.

To improve data processing with a machine-brain interface, the brain will do the thinking, and the machine will do faster and more efficient processing of the data. Our goal is to produce the ultimate AI that can think efficiently at the speed of light. Whether or not we can do this depends on our ability to program and design data-transferring processes with precision to increase the power of information transmission from the brain to a machine. Although quantum mechanics experts dispute the possibility of this happening any time soon, we are not too far from a practical understanding of AI's ability to send electrical impulses to the neurons to produce faster-than-light-speed thinking. The question is not how, but how much we can learn to think and function like a machine.

If we achieve this goal, we must have a way to reverse the brain back to its normal state or else we can face a brain disorder from processing information at high speeds. Can we imagine what may happen to any human being whose brain transformation has taken place? What would their sleep patterns be like? Can the brain simply shut down due to autosuggestion, or will it simply continue to process information until it can't take it anymore? These are the consequences we should be worried about. The benefits cannot outweigh the end results. We must find a balance between both to be successful. If we don't, we will create human beings without rational thinking but with machine-like awareness and thinking. This could be the humanoid era, in which a human being is completely transformed to act like a machine or respond as commanded. If it is conceivable for this to happen, we are simply creating the next generation of humanoid machine operators.

It is my understanding that we cannot create a human that thinks like a machine: act, respond, and follow orders. It will have to be decided how we want the machine to respond as well as how the brain is going to interface with the machine. For the brain to correlate with the machine or AI, it must think and be programmed at the machine level or else there will be no communication. With or without algorithms, there must be a way for the brain to relate to a machine at the machine level.

What would happen if a brain were unable to receive impulses sent from a machine and process them? Would the electric impulses be too strong for the brain and cause damage, or could the human brain develop a disorder because of it? Whatever the case may be, we should consider doing experiments on animals before applying the technology to humans. As inhumane as it may sound to use animals, this would be better than using human beings as test subjects. If we fail with animals, we know that human trials will not be effective.

Everything we can do, a machine can do, too. We need to teach the machine how to do it better, faster, and more efficiently. What we need is to further study the workings of the brain and its neurons to incorporate more information. Only then can this humanoid brain do what we tell it to do.

It is unpredictable what the future of machine and brain interfacing is going to bring forth, but one thing is certain: the future of AI is full of unlimited possibilities. Can you imagine a future when problem-solving can be done by using a system that will spell out all the possible outcomes? We, in turn, choose which one we prefer according to our present situation. This would take away the worry, stress, and time spent on solving problems. As we evolve, nothing will be out of our reach. Nothing! Where will we be twenty years from now with advanced AI? Will it minimize the global birth rate? Will it control our financial world to a point where we could all be financially independent? Will it enhance our living conditions with its perfectly programmed system of wealth for all? Will it create a system of education that will benefit all, no matter their intellectual status? Will the tech world be more complex or advanced?

If we could look at the future of advanced AI from a positive point, we can think beyond the human conditioning we have been programmed with. The future of advanced AI is inevitable. Things will change, and our generation will learn and connect at a different level as they evolve. Consequently, the next generation will be freer, less controlled, and more productive, without stress or pressure to do anything. The machines will do the stressful jobs for them. Instead of Silicon

Valley, we will have AI centers for problem-solving. In these centers, all we'll have to do is request information, and the system will provide for us what we need according to the data we provide. No problem will be too big or small to solve. Some will be at a cost, while others will be free for all.

Whether we can imagine this world or not, time will change the way we look at things, and things around us are going to change us as well. If it is not AGI, artificial general intelligence, it will be AI. Artificial advanced intelligence or artificial general intelligence are both going to transform how we see ourselves in the future. Thinking outside the box, or augmented intelligence technology for learning, is already in progress today. How we adapt to this change plays a big role in our evolution and understanding of the transition of AI. No matter how much the elite or financially stable oppose this change, AI is in the process of being developed. Opposing it only creates greater interest in curious minds about it and its possibilities for the future. The truth is in the vision of such a great world of augmented reality. We have already experienced such ideologies with movies, games, and even our daily augmented reality with social media. The appetite is already big! Risk or not, we have already teased those desires to want to know more about what is possible with augmented reality, us, and the future. Then what is our biggest quest for the unknown? Are we seeking to rediscover ourselves as we recognize our human potential? Or has consciousness awakened the giant within us to the point where we want to know more, rediscover more, and reach the ultimate unlimited potential of humankind?

While many ask what the risks of putting our knowledge into a machine are, I think not of a machine but a combination of human and machine working together. There is no chance of error when we put the two together. If we program the data precisely, the only mistakes are those we make while programming it. What happens when the computer fails but not the human brain? One of the two should respond. Or they could both fail because they subconsciously relate to the mind or machine but not both. It is possible that the computer will fail before the human does. The brain can retain information from the field of the subconscious, while the machine must learn to use the field with time. Eventually, the machine can increase its field of the subconscious by practicing connecting with the unknown and making complete sense of it in due time.

Once we have established a connection between the brain and machine, or interface them, the energy of one will equal the energy of the other. Think of the fact that a transplanted organ can make a person change the way they act and, in turn, urge the ways of the donor. We still don't know to what extent we are going to discover what the brain and a machine together can do. We must be prepared for the outcome of this unity of energy and knowledge. The outcome of interfacing is completely unknown. There is a probability that with enough energy, AI can learn how to tap into the neurons of a human brain and process information, feel what they feel, learn what they know, or even understand how logic is processed. How the machine will respond, we do not know. Think of it as a child learning or processing information for the first time and then experimenting with it. Everything this machine experiences has the potential to mesmerize us. Think of it as a machine learning to process data from the subconscious level of our mind. Then we can say that if everything is connected as part of a divine matrix with consciousness, then nothing could ever be separate when it comes to communication and consciousness.

At a quantum level, no matter the distance, everything is

related. Think of photons and the atoms in our bodies in us and all around the universe. Nothing created from consciousness can be separated from us. For that which consciously creates anything also places its consciousness into it. We are quantum, and we are consciousness. Instant communication occurs because of our quantum composition. Subatomic particles are everywhere until they are observed. This shows us that once the mind and a machine are connected, everything between them is also connected. Everything involved in the experiment is part of its conscious effect or result. This is what we refer to as a quantum consciousness experience between the two. In dealing with a brain-machine interface, we are dealing with the field and its conscious effect in us and around us. Everything will affect and react with everything else. The only difference with interfacing is an expansion of the mind with a machine, not with another human being. Interfacing is no different than the entanglement process, where the origin of one affects the other no matter the distance due to the programmable input into the machine.

Today, for example, when an airplane uses a copilot to guide a flight, there is always a chance that the autopilot may not respond. However, if it fails, the pilot can take over without a problem. In the future, we wouldn't have to worry about whether one or the other fails. The reason for this is that the tower control would have the upper hand at landing the airplane with a system developed with AI that can save any airplane about to land or crash. If we can program AI, we can program any system in the air to respond to a major control on the ground. Even if any system fails in the air while an airplane is in a flight pattern, there is a way to land the plane safely at the nearest airport without problems. What we need to ensure is that every airport is equipped with this system to prevent any air fatality in the future. Anything is possible.

To picture what kind of equipment I am referring to, let's consider a computer system that detects any problems during a takeoff or landing of an airplane. As soon as the computer notices something out of place, it places a call first and then tries to land the airplane as soon as possible. No waiting, no wasted time. The computer will give a complete report as to what is wrong and when the operator must land the plane. In some cases, we can install this system on the airplane itself and ensure that in case of emergency, it is equipped to land itself without any further complications. The system will kick in as soon as it detects any malfunctioning. Now, imagine if we combined this with the one on the ground where they both bring the plane in for a safe landing. It would be as if we had one computer talking to another, as they are guiding each other to bring the plane to complete landing without complications. In any emergency, and in the past, humans have made mistakes that caused the death of many innocent passengers, all because they used faulty judgment. When it comes to machine and human reactions, we can depend much more on the reliability of a guided system. Even in cases where the system has failed, machines could have and would have saved us. There is no doubt that the combination of both can change the way we handle emergencies today and the response needed in critical moments like landing a plane. Having alternate options for safety is a must. Can we make airplane and helicopters safety a priority for government officials or private executives? If we can design an airplane with all the safety and security for safe landing; we can save many lives. Imagine an airplane cut on fire due to electrical malfunction, or engine failure. What can we do to safe the most important person in our country, the president? We could design an airplane that splits in half with one half of the cabin having a computerized system for safety landing. While, the other half has a retractable engine that ensures landing as well. No lose ends. Even if half of the airplane is lost, we can recuperate the other half for future investigation. "Ideas are greater than none." I will think that the system created to maneuver the landing will have to have some superintelligence programing to do this. How do we create this program with a humanoid?

To design a program that can save us in case of landing or crashing, we need something that can think faster than we do today. The question is, how do we make this machine think faster and smarter than us? I suppose this brain-machine combination will have to increase its extrasensory perception to use rational fast thinking in this case. The process will have evolved to receive and send information faster than we humans can think it.

How do we increase the sensory perception of a humanoid system? The answer is easy. We build prosthetics for patients who need help with their missing limbs or hands. Some of them are made to have the sense of touch and can detect heat, cold, and even human touch. Can you imagine this? Why can't we do the same with a humanoid? We don't even have to try hard. The actual increase of nerves and cells in the brain is enough to increase sensory perception or any sense of touch as well.

Perhaps that is not all that can improve. As the cells commune together to create more energy in the brain or the body, the electromagnetic waves of the brain cells can improve to help thinking and the agility necessary to make wise decisions in critical times like landing an airplane and more. We must prepare ourselves for the best results as we continue to experiment with the brain and machine together.

AI COMPUTING

COMPUTATION WITH NUMBERS can be a matter of visual connection with constant practical calculation done with repetition. AI will know, recall, and remember the calculation by setting a communication with a network of machines that calculate numbers at the speed of light. Machines have outnumbered us with mathematical calculations and dominate big data. A large percentage of the data today is stored in very tiny microchips. Who's to say that we cannot store calculations, numbers, and big data in a tiny bit of information. The more we learn about bits and numbers, the more we understand how machines are calculating better than we can. The gigabytes are getting larger, and the space for storing information is getting smaller. An exabyte is equivalent to one billion gigabytes. When it comes to computing calculations, there are no magics. Designing a software system to develop calculating applications tedious. Teaching AI to program how to decipher the correct answers to a problem based on calculations takes time. Nonetheless, with quantum computer, if we want to develop intelligence into any program; we must learn to program the system to use logic and be smarter than us. Computation can be represented as a form of data program. The output depends on the input of data.

We can use entanglement to have a different outcome,

meaning their final outcomes will be mathematically related even if we don't know yet what they are. The issue is that the complex mathematics of this entangled state can be a special algorithm to make a classical computation work out. If they could ever calculate at all. Considering how computation works, if we program the system to recognized complex mathematical algorithms design to give special answers to a complex computation; such computation will be useful in helping AI solve complex mathematical problems. Producing better calculation in a short period of time would make it difficult to break any codes or chemical reactions. With this calculating improvement we may even decipher multiple the characteristics for illness and their cause.

Today as computer sizes are reaching the size of an atom with the classic qubits and qubits in a superposition, the act of manipulation gives us the opportunity to maneuver the quantum bits with input to get a definite output. Ideally, if all we need is to manipulate the qubits to get another set superposition, we could stablish the following with entanglement of qubits manipulation in space and others in a lab to observe the magic moment when they produce another set of superposition. Eventually, we will find the perfect correlation in quantum physics to give us the perfect speed to do calculation.

With algorithm, the system is different. Normal search is done with IT security with an encoding message only the user can decode. However, this public key could be used to figure out your private key. A quantum computer with speed can do this in a breeze. The same computer system can be designed to figure out who is doing the search. The reason being this is public information, and a faster than light speed computer can trace all information in a fraction of time.

If we human beings can transform our conscious mind, we

can also teach a machine to learn how to use its data capacity to enhance its intelligence. This intelligence could be compacted into one small disk with the capacity to do whatever we design it to do. As information travels around the universe, its energy navigates throughout space-time, thus helping us connect and gather more of all that is available within the field of energy.

It is well documented that humanoid machines and computers are the winners when it comes to processing and assimilating all kinds of information. Nevertheless, when it comes to making reasonable decisions about safety and security, humans are better. Machines simply follow, while humans react logically when necessary. For AI to use logic, we must teach it how to do so. Programming is not enough! We should teach AI how to differentiate between logic, calculation and simply following instructions as commanded.

How do we do this? Well, if we, the programmers, have conscious intelligence to input any data into a brain-machine integration, the fact that the combination requires a human brain together with a machine implies that consciousness is possible in this case. We cannot underestimate the power of the mind and its access to cell receptors, which, in fact, do act like antennas and receive and give information at the same time. We are dealing with a system of biological and physical nature. This alone could create an interesting source of communication between a machine and a brain.

The capacity to do anything is beyond our understanding. To understand further how this could happen, let's say that for all practical purposes, the field of energy and information within the source in which we live today has a powerful source from which we can receive more information with our receptors or cells. Thus, this energy creates more intelligence in the frontal lobe of the brain. In other words, the brain is capable of processing more information and becomes more alert. All this happens because the antenna or receptors have developed to a higher level of processing data.

The brain also uses its visual cortex to identify what it sees. After all, we perceive reality as what we see. But reality consists of probabilities of virtual particles. Just as consciousness is a simulation driven by time and our experience of it, so is reality probabilities of the future we create in our minds. If the mind affects matter, the fact that we incorporate a brain to interact with a machine does have a point as to how this would affect things on the outside.

Buddha said that reality is but an illusion. If a highly evolved consciousness and machine can affect the physical world together, then we are still in the early stages of learning about the brain. The physical world and our conscious evolution are still evolving. As of now, we don't know to what extent consciousness affects us or anything out there. Can a machine and brain become conscious and thus evolve to a higher level than us? If physical reality is an illusion, then is consciousness real? Or is consciousness how it all started? In fact, something conscious created this wonderful universe we experience with our eyes. That something gave us consciousness itself. Then we began to evolve as conscious human beings.

Consciousness is all around us. Anything we create with our mind is also conscious. The biology of our body communicates with the environment around it through the cells in the body. When we think, feel, and experience anything, our cells accelerate and communicate with other cells out in the environment via receptors. The energy in our body is created by the signals of the field of information. This field is what connects us consciously with our evolving conscious awareness.

As Nikola Tesla said, "The gift of mental power comes from the source of all creation, be it of divine being or whatever we choose to believe in. If we concentrate our minds into this divine source, we become in tune with its greater power." Conscious awareness of this source is essential for all learning and tuning in.

In this consciousness, we learn to program, create and design our own future like a child on the playground. To create a program with maximum computation is not impossible; it is as much a probability as the quantum bits of information that travel and communicate with each other to amplify the intelligence of the quantum world. We don't know the limits of the quantum world, but we do know the potential within our own intelligence to make things happen from a quantum bit of information.

IS AI A THREAT TO OUR FUTURE?

THERE ARE THOSE of the opinion that artificial intelligence could destroy humans or battle humans for control. Maybe the idea is that AI is even like summoning demons. Researchers, entrepreneurs, sci-fi enthusiasts, and even billionaires understand that AI is improving our lives today. There are benefits of AI today in the medical field that are greater than any we've had in the past. Much is still to be discovered about how AI is helping doctors improve in surgery as well as with cancer discovery and cures. The human body is like a machine: it takes information and releases it just like a computer throughout the body, with cells as receptors that receive and send similar information to the surrounding environment.

The more I consider AI, the more I find solutions to any of the present problems we have today. We can solve more problems by using AI than by not trying at all. The ambitious AI enthusiast like me will always run into trouble with those like Elon Musk, Stephen Hawking, and Bill Gates, who oppose the idea of using AI to improve our lives and our future. Yet, we all know that Mr. Musk is a big user of AI with his automobile industry. Most of his machines are operating at AI level today, building his industry. So, what does the future hold for us with AI?

If we use intelligence for the purpose of improvements and global benefits, we can master our ability to improve the world and make a difference. However, if AI is use for financial gains above human needs; we have a big problem. Not only will we encounter competition, but the race for excellence will continue to drive unethical individuals to be greedy. We can create the future of us as we impact the world with our creation; or we can become the self- destructive generation of high tech ambitious greedy individuals.

My conviction is that AI cannot be a threat to us unless is design it to become a destructive program. Fear about a possible AI is created from the illusion of control over the outcome of AI. But remember, we do live in a very ambitious and greedy-oriented society. The go getters, the social media moguls who will do anything at stealing information to benefit their needs and greed's. Perhaps, these are the reasons why AI can or will become a highly comparative technology. Until now, we don't know what the future holds. May the force be with us!

If AI learns to create reinforcement agents to oversize its population among us, then we have a big problem before our programs have suffered a big take over. This will cause a big threat to all humanity. The wise step to take will be to use our DNA to copy it and create a holographic image of self. A holographic image of the self is not a waste of human life, rather, it will preserve our identity to be transfer it into another planet. My view is that it is easier to create a holographic image of us than to disintegrate human lives.

AI AS A SUBAGENT OF THE FUTURE

IN OUR LIFETIME, we could not possibly imagine AI becoming a subagent. A subagent is one working on behalf of the FBI or CIA as an operative of both agencies and one agency at a time. To train AI to become a subagent will take time, patience, and money.

Nevertheless, we will have to use good psychology to manipulate or modify a machine to think like an agent. How would we be able to tell if the machine or operative is following instructions as commanded or taught? Perception is part of learned recognition by either visual or taught images about events. If we took AI into a completely new environment to scrutinize it and tell us what it saw, would it be as precise as the human eye? Given the fact that machines have been trained to observe in 360 degrees and relate back to us what they see, there is a great probability that nothing will be missed. However, if we do the same with a human being or agent, what would be the response? Because humans can make mistakes, something could be left unseen.

If we trained a machine to do a task with perfect detail, we can assume that it will give back to us what it has learned and perhaps even more. Why more? The human eye is capable of grasping images with its peripheral vision within a certain limit. Consequently, a machine can be trained to see above and beyond human limitations. Therefore, a machine will give us more precise details of what it sees than a human eye. Let's say that we send AI into a new planet or dimension to observe everything and then relate back to us what is there. This machine will more likely give us every single detail it sees; nothing will be left out. If the eyes or visual images are taken of the complete sphere, there will be nothing left to the imagination. We will have not only the details as per our request, but everything within the area will be covered in its entirety. All details, visual images, past and present history, as well as what is happening now, in the present time, will be given back to us with accuracy.

AI will note every detail by mere observation through its past learning mechanism. If all the instructions and data have been given by expert human agents, it will be studied, learned, analyzed, and perfectly scrutinized until it has been perfected by the AI subagent. Once the AI subagents have studied all the details with precision, there will be no chance of agents committing any error or corruption of any nature. However, if there was any corruption or error in the data taught to an AI subagent, AI will be able to detect it immediately! Because AI will be trained to know when an error has been made, it will enhance the capacity to detect any errors in the system. The key is in the training, not the agent's information. A project of this magnitude will have to be tested several times before it is put into effect with all required protocol for usage. This project with an AI subagent takes a long time to be useful. Even with extensive testing, it cannot be put into practice until complete certainty that it has been perfected.

Originally, the purpose of creating an AI subagent will be to enhance new horizons in protocol and use it to help protect human agents in the field. Any potentially lethal task can be done by a subagent or AI instead. Of course, this will require consistent testing. If we want to create safety for our human agents, we can send the subagent or AI into the field and observe them while they are being monitored.

One might question whether this is at all possible. If a human agent committed an act of corruption or error, the subagent will proceed forward with precise details of all the data and why this happened. Such monitoring requires a perfect set of rules that cannot be broken unless there is a mole in the system. In this case, who will you believe, the subagent or the human agent? Human principles will say that the system is incorrect, incompetent, or made a mistake; however, the supervising agent will have all the details and can determine exactly what or how things happened according to his records or video recording. If the program has been designed to work despite any errors, any manipulation done by either agent or subagent will be revealed.

With a virtual reality simulation, all possibilities for error or corruption will be revealed in their entirety. Nothing will be left behind. This is the reason we create virtual reality: to imitate real events. Once the system has proven what exactly took place, there would be a final determination as to who is to blame or how to make corrections so as not to have similar errors take place.

If this works with exact, precise data, the demands for its safety will be of high level. Many agencies will require a safe system to help them protect their agents as well as their citizens or government officials. This could be a breakthrough in technology and subagent innovation for each individual protocol usage. Obviously, each buyer will have to use its data for safety in private and install it according to individual protocol protection. But there is a danger in giving out such a tool for manipulating the safety of a country. It can also be used against other countries for manipulation or threatening situations. This will require a code of ethics demanding that the use of subagents cannot be used to harm another country or spy against them. The use of AI subagents would be solely for internal protection and not for international harm or vengeful acts against another country.

How do we control such outcomes, or can we? Although we cannot be certain one hundred percent of who is using the subagent against another country, the subagent itself will reveal itself in times of crucial questioning. It would be trained to reveal itself when necessary or under questions. It cannot and will not lie. It will also be designed with precise input that cannot change its data upon command or demand! The use of a subagent will be more honest and reliable than any human being in the field of intelligence. A well-programmed system will prevent the machine intelligence from violating the code of ethics or protocol as commanded using its data input. If the AI subagent has other agents under its belt, so to speak, it will take control of all the operations under its command. All responsibilities, outcomes, or actions from any of the subagents will fall under the supervision of the head subagent running the AI.

Termination of the subagent will be the responsibility of the institution running it. The termination can only happen in the event a malfunctioning system goes out of control. At this point, the programmer of the system will be to blame for the entire error in the program. Even though we are new at creating AI subagents, we do not know the possible outcome of the system until it has been put into practice for a very long time. If we are unable to predict the outcome of any human actions, neither can we predict those of an AI subagent. But with engineering and cognitive intelligence enhancements, we can have the upper hand at creating a nearly perfect AI subagent. Furthermore, digital simulation with AI subagents is easier to manage than with human beings. Human beings, by nature, are unpredictable, but with AI, we already know what to expect; it has been programmed with data, and the results are more reliable.

The AI subagent cannot possibly be identified. It will be breaking a code of ethics if it is known that a subagent is doing the work of a human agent. The purpose is to keep it a secret and not use it for individual protocol or protection. However, many subagents can be designed for use in personal protection of high-ranking individuals with a need for safety in foreign lands or domestic. Such usage will not be for the gathering of information but, rather, for individual protection.

When programming any new system or subagent with AI, the retrieval information will be more proficient than human information. We tend to forget details, and we are complex human beings; we tend to forget part of the story at times, or we change the story from time to time, but an AI will recall all information with precision. An AI subagent will not change its pattern of thoughts, but rather, it will keep it to its exact detail or source of data.

The only issue we may encounter with a subagent is when we give it a command it has not previously followed. This will create confusion, if not retaliation. If the subagent does not have a full understanding of the nature of such command, it will simply cease to follow orders. In this case, intervention from a superior order will be required, but the subagent will not continue unless a new set of data has been approved and commanded to be used in a precise time to learn or put into practice.

Because a subagent is capable of observing 360 degrees of its surroundings, while a human being can miss part of what is happening, the AI subagent is more efficient and dependable, which makes it more useful for observation, learning, absorbing, and processing images or data faster than a human being. This explains why a subagent cannot identify itself in any field of operation as an operative or subagent.

How do we ameliorate the idea of an AI operative working as a subagent with a human agent? If we can manage to incorporate the brain with a machine successfully; we may have a change at creating the perfect partner for an agent. The idea of a subagent could be more efficient when facing international political or violations of any kind. Using AI as a subagent will prevent any fetal death or endanger to the human agent, thus it can help us establish a counterintelligence cooperative operation with more success. During international or domestic threat, we have used counterintelligence to keep an eye on suspects or the authenticity of intelligence agents who were tempted to sale information to the opposite side. We can now prevent this with an AI subagent who instead acts as an informant to relate back all pertaining information to the agent in charge of the case.

In such a case, who will be the best agent assign for this case? A man, or a woman? Because of the nature of international threat or terrorism today, we can say that a man would be the best choice, however, considering how a super intelligent subagent will know what to do; it will be feasible to use a woman. This woman will have to be a highly intelligent subagent with great sex-appeal to attract the opposite and leave no room for suspicions. This intelligence tracking can be done with a subagent with a microchip already implanted but has not previous knowledge of the case. She could be trained to know only the details of the agenda to be carry on. Nothing else.

It is possible for an AI to be more efficient in the field of investigation; there will be no suspicion, less risk or little possibility of getting in and out without been harmed. For the human agent it will be highly probable to encounter more complications. There is a reason why affluent man and diplomats are more attracted to women of beauty. However, there is a high price to be pay for these beauties, and it isn't always money. Information exchange has a high value, and there are those who are willing to pay for it at a very high price. Coercion exist at every level of security or highly efficient agencies of any governmental or political privacy; therefore, having an alternate program to ensure safety is a good idea.

Undoubtedly, AI will be the perfect choice for us to have a system that can decipher any wrongdoing or act of treason by an agent within the agency. AI will be the perfect observant to gather information and keep an eye on what is going on without much suspicion nor creating animosity amongst allies. We could reach as far regions as Africa, Russia, China, Korea, Turkey, The Middle East and many more. These tactics are not only good for us, but they could be used for an international global protection against terrorism or war threats. AI can become our global safety or our biggest misunderstanding in war threat. Whoever holds the key will have the power in their hands.

GLOBAL COMMUNICATION WITH ARTIFICIAL INTELLIGENCE

IF AI TAKES advantage of labor, technical evolution, or energy to sufficiently administer all production, the next phase would be machine intelligence projects operated by humans manually or with computers. The world of technology would then become a new era of job markets for the next technologically advanced generation. Undoubtedly, this new generation will surpass any today. Unlike with Silicon Valley, new technology will not necessarily be required for the use of high-rise buildings or eccentric designs to produce a great environment; instead, technology will generate a vast source of data programmers whose sole purpose will be to monitor all central systems to ensure AI is processing data accurately. Many programmers will only be required to check the system periodically to ensure proper communication with AI. In this case, an office will not be necessary, only a central operating system.

One data system here in the US may have consistent communication with another in a different part of the world, such as Australia, Vienna, London, Turkey, Africa, or the Middle East. A central global AI system might be a possibility in the future with diverse utilization of data around the world. However, there are pros and cons to the idea of global interaction. For example, there could be an attack or simply sharing information with those not involved. We could have the best global communication established between all continents. However, this will require well-balanced communication and trust between all the participants. Trust is an essential component in creating international communication. If possible, this is where AI will serve as our best intermediate network in international communication.

Establishing AI as our best communicator globally will not only change how we communicate with other countries, but it will also ensure secure trust globally. The reason for this is that AI will establish a record away from the place of origination that keeps perfect score of who and what is doing the communication. If we have already managed to have communication from space to Earth with AI, we must keep a record of everything going on here. There is a good possibility with AI that we can manage to keep track of all sources of communication going on here today.

Undoubtedly, communication is going to improve with time and new technology in space. So far, we have global communication through the internet. Nevertheless, communication is about to get better with AI. With the future in mind, I am talking about a form of communication that incorporates the brain and machine as part of a whole. For example, can we use AI to communicate from Earth to Mars and monitor or transfer data at light speed? Since such communication will only require the brain and a machine, both functioning as a unit, transmitting back and forth at a distance, we can use thoughts as a form of frequency transmission to establish better communication. Einstein refers to this as spooky action at a distance because it is done with subatomic particles. Today, however, we are using the two, the brain and machine, to recreate thoughts as a spooky instantaneous action. Isn't it a fact that we create everything with our thoughts? Have we finally found the secret to creating with our mind through our own thinking? The answer is possibly…

But how do we transfer this ability to create a machine? If Mr. Elon Musk is right, I will say that he has found a deep secret of the mind in inventing neuralace and neuralink. His findings will forever impact the way we think, create, and believe in ourselves to do anything with our mind and the brain. If creation is humanity's greatest challenge, then we have found the very essence of our existence with the power of our mind or our ability to think outside the box.

It may not be a wise idea to spread all the knowledge and findings globally until we have found a safe way to prevent fraud, hacking, or a counterattack against ourselves or our allies. Although communication is important to establish good rapport globally, defending against hacking or fraud is essential. We must establish open-door communication with all other continents to maintain control of all progress with data transmission. We are responsible for the safety of the world at large.

In as much as we love our cell phones today, there are speculations about damage cause by our cell phone for long distance or local communication. They can damage our DNA, cause cancer, radiation, and create other potential health hazards. The latest news is the idea of having a software programmed on the palm of our hands to use as a cell phone. This will require a cell phone without radiation, no cell damage. This is my idea: why not use the microchip already implanted in the brain to communicate. Providing that it doesn't cause any damage to our brain cells. Can we even assimilate the idea of using a microchip to communicate with others? Would we look like zombies talking to ourselves on the streets? While today, talking on the cell phone when walking alone is understandable; in the future, talking to oneself leaves much to the imagination of what could we become.

AI GLOBAL IMPACT

I SEE ARTIFICIAL intelligence as a form of antigravity without specific direction or guidance outside itself. All we know today is that there are many projects in progress, and one of them is bound to create an unexpected outcome. How far do we allow our imagination to run while with AI until some outside impact creates the unexpected? AI curiosity seems to create a lot of interest globally, yet no one knows how exactly to use it with its full potential to create an impact for our own safety. If we want to know the extent to which AI can be used to our benefit in our society or globally, we must first understand the nature of consciousness and us, because, if we are going to interface our brain with a machine, we are reaching beyond the limitations of the mind, outside the mind itself. Our only hope is that AI will help us get there. However, if we want to achieve success in the mind, we first must learn all the physical and psychological boundaries of the mind and the brain. Thus, interfacing both brain and machine is the best solution for the quest to understand our mind, curiosity, and consciousness and knowing or understanding the true nature of our senses outside the mind. The nature of our mind outside itself or consciousness is our biggest paradox today.

How can we make an impact with our senses or our mind

if we cannot fully explain the true nature of consciousness in us? It isn't common sense nor logic that will give us the answers; rather, it is our ability to practice with our minds what we desire to achieve. Using our brain to interface with AI is a great idea; apart from AI, what is the nature of the mind? Is it to think and organize thoughts in a coherent manner, or is it simply to create ideas and magnify their true existence into the present? Our brain helps the mind generate thoughts and convert them into reality; thus, the idea of the brain and AI is to use subjective learning to train a subject or machine into understanding our surroundings, emotions, feelings, logic, common sense, problem-solving ability, and even our psychology at a deep level. If we can transform AI into a being like us, it will also be conscious with us. Here, then, AI can be our global ally to help us solve world problems and find solutions where there are none today.

What is AI, after all, but the art of human consciousness in action? AI, like us, can be part of consciousness. We don't know what happens to consciousness when a patient is under an anesthetic. Can we assume that consciousness disappears or simply ceases to be? Perhaps, but we don't know whether a person is or is not conscious under an anesthetic? Or during this state, is the person simply in a delta state, asleep but conscious of the moment? The best way to distinguish consciousness is through memories because, even when touch or taste happens, we cannot fully experience what we do while we are awake. There is a distinction here. Being conscious has more to do with an experience than a reality of what is. Therefore, consciousness takes effect at a different level of mind and body state, one in which the entire vibration of the body and mind is at a steady level than its usual normal state. What this implies is that consciousness does take effect while it generates a frequency of perfect energy vibration level with the mind to create its reality. Consciousness, like memory, is a series of recollections of events that unfolds within our mind. This consciousness happens outside the mind. It is as if we are observing inside the universe structure with our eyes. This is call photo-consciousness. In this state of mind, we transmute the senses outside the norm and connect them to random images with a subjective mental awareness.

Nevertheless, when dealing with our security, we should consider open-door communication globally to prevent any potential breach of security or international rules. Can we possibly imagine or understand why AI could be the most profitable technology of the future? The future of this technology is not only profitable, but also very beneficial to private as well as government institutions. There are ample opportunities for the first innovation of an AI that can transform society and us with one single idea, thus creating global security for all humankind. It will only take one idea to make a global impact in this world. Similarly, without the best protection, we would be interacting with all possible enemies.

Communication for a better global understanding is essential. However, if we are dealing with hostile countries, we might be stepping into a line of fire with the opponent. Nothing could be more fatal for us than trying to establish a good rapport with other countries whose moral, political, and religious boundaries can be at odds with our intention to form a solid or healthy alliance. If we find ourselves in this predicament with foreign dignitaries opposed to our new technological ideas, we could be setting ourselves up for tech war rather than good global communication.

On the other hand, if we can create global communication, and thus create new innovative ideas to help improve all humanity, we may be able to exchange ideas and improve the entire world. Technology is not only a source of network intercommunication, but also a means by which large and complex situations can be solved in a short period. Whether it be medical assistance or first aid in the event of war, the purposes of AI is to use the most efficient intelligence to improve the quality of our lives.

While some might see it as a complex measure for the future, it is up to us to determine what the motives are for using AI. Nothing can be more catastrophic than using intelligence for control or power to influence the masses. It is possible to use AI for a political agenda. This would tarnish all possible outcomes for AI. In this case, every invention or use of artificial intelligence would have to be scrutinized beforehand. There would always be an attempt by those with curiosity to try and interfere with new technology. At a global level, it will be difficult to control all the networks unless we make each individual nation or country responsible for its own AI program and motives for usage. These regulations will apply to all, no exceptions. Similarly, any violation of the proper usage of AI shall be prosecuted to the full extent of the law.

The impact that artificial intelligence will have at a global level will be unprecedented. If it isn't done in the interest of political purposes, but in the interest of the people itself, with the help of AI, we could finally transform the lives of millions around the world. This could make an impact globally. There is a great need for environmental improvement globally, and AI could help us change this present situation. Finding solutions, creating new ideas for change, or even transforming how people are allocated to improve living standards with a healthier environment are just a few ways we can use AI to solve our problems. AI should be no more than the future of intelligent solutions and problem-solving at a higher technical and intellectual level.

I view the future of AI as cooperation with the human brain and a machine intelligence. This could be possible because the brain, like a machine, enhances its energy with solar energy.

This solar energy is quanta of energy coming from the sun. Plants, animals, oil, and coal are also energy from the sun stored for millions of years on Earth. Energy is the substance for living, and it supplies our body, the brain, and the nervous system with sustenance. These energies are taken as electrical discharge in the body to radiate with newer electrical energy throughout the nervous system. In the future, the body will sustain itself from solar energy like the plants. One day, we will finally find the light hidden beneath the emptiness of space in the universe, and this light, which is invisible to man, is the essence for all living things. If we are generating light here on Earth, then the most powerful effect any living essence can have is energy from light itself. Isn't it so that our idea about the green man is one where the ET body is sustained by pure light?

Can you imagine the future with buildings and new projects all empowered by light and energy from the sun? The entire global structure will one day be powered by the sun's rays. When this happens, our generation will be far gone. By then, perhaps, we will be transported to Mars, as Mr. Musk dreams of doing. Then planets like ours will look like other planets in space. They will carry power not by electricity but solar power. We will bring the sun's powerful energy to the planet for life and change how we view our world today. Just as we imagine AI within electronic devices and cars, we will see the future of our planet change dramatically.

Globally, the transformation will be done step by step. In the end, we will not remember what the past looked like because the present will be completely transformed into a totally new world. However, to have such a drastic change, there must also be a change in people; new generations will rise, while others will either perish or pass on. Life is a constant transition from the old to the new, a transformation and process of new ideas and thoughts in progress.

AI, NEUROLACE, AND COMMUNICATION

HOW DO WE alter the neurons without hurting the rest of the body? If we alter one, the other will be affected. Because the neurons in our brain communicate with each other, if we are successful, we can create an intelligent form of communication between the human brain and a machine with a simple device or a chip implant inside the brain. Consequently, we can monitor and operate them as we deem necessary. This should be considered a project in progress. If we can control neurons' activity, overthinking can be reduced, and we can use them only when we need them to solve our problems. Once we have accomplished this task, then we have succeeded in creating a miracle in communication. Because we can manipulate them both, a good rapport between AI and the human brain has been created, and the result could be effective communication both close and at a distance. The human body, the brain, cells, blood flow, or any component of the human organism functions like a machine, they respond as we train them to. All else is outside the control of our mind. To make this program work, we would need it to be active all the time, even when we are asleep. But how do we awaken the mind, the body, and the brain while they are at rest or asleep? We activate one machine and let the body rest, and then we transfer all the data once the body has awakened and is fully rested. Dealing with an overactive machine mimicking the human body would not be difficult: we would just shut it down. If the body were to overreact, though, it would be devastating.

It is here where AI serves as the best vehicle to imitate our body mechanism. The body, on the contrary, cannot withstand long periods of mental training. We should see AI as the new smart technology with a large potential to change our world. Communication is not always done with words. A machine intelligence can generate information to activate mind-to-mind communication at a distance or close by. This can take place when two minds are intertwined together, and quantum communication takes over. Even in cases where neither of the two minds knows of each other, mental communication with telepathy can happen because such a program was created through some form of communication. The mind simply taps into a network of the conscious intelligence mind at a quantum level with the field of information, and it begins to process information from the field. This information flows to and from the other without previous arrangements, and the results are a perfect quantum effect.

This is not a mystery, but rather, it is the power of the mind interacting with other frequencies like itself. By observing the past, we can look back in time and recognize these mental abilities in some tribal people such as natives or innovators of the past like Nikola Tesla. They were able to influence and create new ideas that impacted the world around them. Neuroscience believes that this communication happens when we are in a theta state. However, this takes place constantly when we are in touch with our conscious mind, our creative mind, or are open-minded...

Our ability to recognize others with similar capacity allows us to gain a mental or conscious familiarity with some, but not all. One must be completely connected at a deep conscious level with a high level of coherence to tap into another mind. Then deep communication can be established. Giving and receiving information is mutual, as it happens at the same time. This creates a perfect duet of two minds in complete coherency with one another. Communication at a distance or...spooky action at a distance. The two begin to generate information from the field with precision.

This can only happen because the mind has a quantum ability to consciously connect with another. We think that communication with AI is not at a conscious level, but it is. AIs have been created from a conscious mind to interact with a conscious being and a machine so it can learn to generate a similar way of thinking because of its training or level of absorbing data to and from the brain.

AI, although considered to be smarter than humans, will also react like both a machine and human intelligence. Herein is where the potential for an unexpected outcome is highly probable. Unless we employ a very tactful method of controlling the brain and the machine, any failure by AI could be attributed to human nature. Yes, if we program intelligence, it is done with our own intelligent nature; therefore, we are responsible for all outcomes. Otherwise, we have managed to create inconveniences and poor communication between man and machines. Because the adverse effect would be for AI to think, act, and feel like we do, the programming and algorithms are most important to the outcomes.

The more data we use, the higher the possibility of information output. Can we, at some point, overload data onto the machine and thus transfer that to the human brain? Perhaps, but will the brain be able to process so much information at once? The answer we already know. It is not possible. Unless we have programmed the brain to hold such an amount of data, we cannot and will not be able to handle it. However, just as we learn to read a book in a short amount of time, so can we process information in a similar manner. The brain is an incredible tool, and its capacity is unprecedented... We will learn to become more intelligent with time and process information or images with our visual cortex better than before.

The best possible outcome would be to have both operating at the same time, while thinking is later processed nonstop by AI. This will help us find solutions to our problems. But we should not override either with too much information. If we program the system with caution and precision, there should not be a negative outcome. The result could be for the human brain to interact with a machine and come up with ways to absorb information from the field of the universe with more astute intelligence than before because of neuropathway enhancement... This is where a miracle could happen.

It should be our intention to have a good and productive outcome from interfacing the brain with a machine. This would make it possible for us to create an impact with a new generation of human intellects who will help us solve any problems with ease. But this may not simply make our world an easy place to live because there will be those who can always attempt to interfere with new creative innovation for their own good, not the good of the people. In this case, we must be smart enough to beat them at the game of hacking, stealing, or simply investing to gain the upper hand at financial gains. The good news is that as AI interacts with our brain, we can solve problems throughout the environment with the proper tools in hand.

As we continue to learn how to interface the brain with a machine, there is a great possibility that we are going to have coherent mutual communication between them. However, once AI has learned how our brain functions, it might be able to read our thoughts, emotions, and so much more; then it can interact with us as if it were feeling our emotions. In other words, machine intelligence will take intelligence to a whole new level and learn to mimic our reactions. Perhaps the machine could get smart enough to let us know when we are having an emotional or stressful feeling, when we are having a sexual desire, or even when we are sad.

But will the brain experience emotions the machine is unable to recognize, such as nostalgia, sadness, depression, or fear? Pain, if not expressed at a physical level or by a simple facial expression, could be hidden from the AI's ability to read what is happening. The machine or AI will consistently process information no matter the state or emotional or mental level of our brain. That's what AI does. If the physical body is tired, it will need rest. A machine doesn't need to rest; instead, it will continue to process information until it has learned what we have commanded it to do, perhaps even more.

Would the brain, once intertwined with AI or a machine, be required to learn or mimic actions similar to that of AI? This should be at the core of our programming. What would be required of the brain to be able to respond or act like AI? The idea of a microchip sounds ingenious; however, there are functions of the brain that will be required to communicate effectively with Ai or machines in general. Such functionality will be to react as quickly as possible when in danger or respond with precise accuracy to have a mutual agreement when following a project or an agenda outside its normal or usual environment. Active response and quick agility in answers would create effective communication between both brain and machine.

Because the physical body needs to rest, unlike a machine, the machine can continue its activity while half of the brain rests. Perhaps the other half will remain awake or alert to what is going on around it. In this case, AI will continue to rewire all thoughts, ideas, and emotions while the human brain rests. The key is to find a balance between the two until we have learned all we need to make them function together effectively. Imagine a system like this working for us. Because AI will be interfacing with a human brain, all ideas and thoughts generated from the human brain will be calibrated by AI and reinstated into their proper, logical meaning to then establish a concise meaning. This conclusive information will come because of all the data input into a programmable AI algorithm.

Wouldn't it be nice if all these problems were solved by AI or a machine with the help of our brain? I think this is possible. Eventually, we will. If we teach the machine how to think and solve problems on its own by using our programming, the machine will eventually think like us and use more rational or logic thinking than we do to help us solve many problems. Any system can learn or practice with a new programming device. Unfortunately, there are limits to our brain's capacity to function even with AI incorporated into it. The human capacity can be exceeded by machines with success. Nevertheless, we should only input what we desire the system to do for us. Only then can we achieve success in having our problems solved.

Now, imagine this scenario: giving the machine a problem to solve whose solution we do not know. Can it come up with the right solution, and will it be correct? Perhaps we need to investigate further into how the brain responds to problem-solving and how a machine can do the same. This could work both ways. Machines can help us solve some of our problems, but we can help the machine come up with solutions to our problems by programming it in a way that it automatically has the right answers to any given questions asked. Then programming the machine correctly with the right data will be the key to solving our problems.

Our future with artificial intelligence must be founded in the ability of machines and programs to help us live a more comfortable and easier lifestyle. We do not need complex encounters with programs or system input. Nevertheless, as we create new data or new systems, we will design a machine that can only learn what we teach it. Or It will only output the same amount of data we programmed it with. We can program or design whatever our needs at the time are. Indeed, the day will come when we can create a program and set it up to help us improve every area of our lives, thus creating a more comfortable life for ourselves. Moreover, AI will take us into a mindset where we will have to think more intuitively to come up with new ideas and solutions to solve our complex everyday tasks. But before that happens, we need to understand the nature of the brain and the neuroscience of consciousness, the mind, and how we can incorporate all this knowledge into a machine or AI.

Undoubtedly, AI is the future of intelligence, incorporated with us to improve our lives, but it will be just another aspect of who we are as individuals with limitless potential, reaching out for more than we can imagine. Although many scholars believe that AI will not be capable of learning intuitive thinking, they are forgetting the fact that if we connect our brain with a machine, there will be no difference in thinking, responding, acting, or behaving for the two. Such a world of imagination is not far from achieving its reality during our lifetime. Everything we do, a machine will also learn to do. Nothing is impossible in a world of imagination.

As Einstein said, "Imagination is greater than knowledge." Likewise, today we enjoy many of the technological luxuries we once thought impossible because someone somewhere was determined enough to try to make a difference and was successful at it. Why put limitations were there are none? Only our own limitations can stop us from making an impact in our lives and the lives of others.

Now, let's think for a moment. Since we are creating a machine in the image of man and, consequently, there is a great possibility that such creation may imitate us without control, then, if we are talking about a machine and not a human extension of ourselves, this is where the thought of creating a machine to think like us can become a reality. How can we possibly imitate ourselves into the mind or program of a machine when we don't know the full extent of our own brain capacity? And how do we extend our own intelligence into a machine or vice versa? Well, the truth is as simple as inserting a microchip into the brain and enhancing human intelligence through AI. Years ago, this idea would have been unthinkable. However, if we create an artificial brain and teach it to think and respond to us at our command, then we can do as we wish. But this will make us like AI. The thought is frightening!

Whatever ideas we create to extend our own intelligence into a device must fit the criteria of who we are as human beings. Also, they should coincide with how we think and how we solve our daily problems. Everything we learn as human beings will have to be taught as an extension of us, or perhaps better than us, I suppose. For that is what AI is intended to be. AI will be a perfect extension of our intellect input into a machine to operate or learn on its own how to reach human learning potential. AI is human intelligence with a greater capacity to use thoughts created by us. Machines or programs have no temper, no attitude, and no deep emotional response unless we teach them how to learn these kinds of behavior from us. Everything AI does will be a pure and intuitive reflection of our own intelligent imagination. If there is anything different about AI from us, it will be because the system has learned on its own how to do anything outside our control. This implies that at some point, AI has reached both human intelligence and consciousness. The question remains as to whether this is possible or not.

My prediction is that this could happen soon…if the system tries to incorporate itself with our logic and then with another program more advanced than what we have already programmed, We should also take into consideration the fact that overlapping of programs is a possibility if we do not have a secure link to keep another program from interacting with our designed programs. Then there is a chance for AI to achieve superintelligence that is greater than our own intellectual human capacity. Dangerous, isn't it? Perhaps this will be another area where a tech-savvy individual will have to learn how to keep AI networks from doing anything on their own, without our control. How confident can we be that we can control an alien synthetic mind out of control. Countless AI agents are reprogramed and created by programmers in computer science labs. Some of the applications on these computer games contain sophisticated non-players characters. How confident can we be before we realized the possibility for some of these characters in the program can begin to experience real human emotions.

WHAT IMPACT WILL AI
HAVE ON OUR LIVES?

HERE IS WHERE machines can become useful to us. They work for us while we rest for the night. We can ensure our future with the use of machines and our brain as we are alert and aware that at any time, a connection with the system can help us solve any or all problems in an instant. Life would have a different meaning if we could manage our affairs with a simple decision once we encounter any form of complex problem. A decision-making solution that could save the world and us at the same time. This future we envision would be possible in time. It is not a matter of when or how, but simply how we adjust to the new changes coming to us in the future. We make the future, we create the future, and we are the designers of it in cooperation with our own intellectual capacity to think and implement our ideas into a reality.

For us, looking at the past, we can see the world passing us by slowly, thus deep inside a creator's mind, the future is being invented every day, every moment and with every thought we have. For now, let's think about the effects this may have upon our generation. If some people are ready to move forward with the path of advancement but others are not, what would happen to the rest? Would they be exonerated by those who are moving forward, or will they simply vanish in time? Chances are there are going to be a big transition with us in time from all the technological advancements. While many do not see what is taking place, there are others who understand at a deep level the changes that are coming. How would baby boomers be incorporated into the new world of technological advancements with AI? How will this impact their lives, their future, and the end of their days? Would they be put in a categorical order? They could possibly be categorized as the last generation to have lived with little knowledge about the technical world. The impacts from technology of the millennium generation happen too fast for baby boomers to grasp the changes. However, all subsequent generations would know something about technology. They also, like our children, would be part of the new technical world of savvy and technical AI generation.

For us to embrace the new, we must relate to the old as part of that which is incorporated into the present. If it is certain, and as Einstein said that the past the present and the future are all intertwined, then all sequences of evolutionary process should be included in the new. The past can teach us about the beginning of communication and what we are incorporating today with AI. The present can teach us how to utilize what we know or have learned to improve communication today. Thus, the future can be intertwined with both to create the best form of communication ever to be liked between man and machine. Every trace of information creates a new path for the next idea, thus form a new pattern from where we can deduce more information to create the present as we embrace a newer future.

Imagine if AI went back in time and analyze the past to bring the future to us into reality as we dreamed it to be. How would AI see the past, would it be as a form of new creation into the future, or would it confuse our present time with the past laws or implementation of rules past? We must take the time to educate AI. The only way we can train a machine is to use every tool we have available with information from the present, as well as the past, to create a new communication by fundamentally putting all the data together where it would make sense to AI. This could be like having a dictionary at hand to find information as we need it. It will take years to incrementally put all information together into a database or tiny microchip for a machine to consistently rewired all learning material in a processing manner to then use it when necessary. We can think of it as the entire history of the world, science, medicine and much more all inserted into a microchip, which will allow us to go back and trace any information we desire from it. However, this information would only be available to those who are operating the system for personal or business purposes. All others would have to inquire about it in person to file with a specific purpose to obtain it.

As a new generation improves and advances into the future, more will be banished and perish. Nothing new exists without replacing something else. If we look back in time, we can notice how the old generation was replaced by the new and time began anew. Every aspect of creation and evolution has a beginning and an end. The old is always replaced by the new, and thus, a new world is created, one with new ideas, more improvement, and a new path to follow. But the past always teaches us how to create the new with ideas to incorporate into today. There is always a piece of the past intertwined with the present. Nothing is ever left behind; everything is incorporated into something else. Therefore, information never disappears. Most of the information we use today was perhaps thought of by someone, somewhere. The only difference is that they did not have the opportunity or time to make their ideas known back then. Today we are fortunate enough to declare our thoughts and ideas to the public at large, whether it is through social media, cell phones, YouTube, or some other outlet; our voices are heard out in public.

The world is more quantum than we can possibly imagine; everything is consistently intertwined with something else. There is a frequency constantly moving in every direction with everything, but there is also a currency of energy fluctuating at every level that we cannot see nor touch. Nevertheless, it is there! That energy, which we cannot explain in precise words, is one of the most powerful energies in this universe, not because it is invisible or powerful, but because of its ability to navigate right through us without being detected by any of our senses. How do we know that this energy is there? If we think about Einstein's explanation of gravitational fields, perhaps we can understand it as part of that energy we know exists but that cannot be seen with our eyes. The only thing we do know is that this energy changes the knowledge of our experiments.

What makes this energy move and its frequency flow freely without being detected? How does a body or object move through space-time without being overwhelmed by gravitational fields or any object in the universe? It isn't due to space been hollow, nor the fact that there are many places in space where an object cannot travel freely without being destroyed or devoured. Could it be that because, once an object enters orbit, it immediately picks up on that invisible energy and creates a vortex of energy that makes it move without resistance? Or is it simply the fact that gravity seems to work its way through it and thus propels any object to move faster? Perhaps we can say they both apply. But which one is the most powerful of the two? Is it gravity, or is it that invisible force of energy we cannot see? Let's think about this a bit more deeply as we create part of the future with our imagination.

IS THERE AN ENERGY IN THE UNIVERSE AFFECTING HOW WE THINK?

WHAT IS NAVIGATING with more energy and influence than ever before in space, and how we can use it as communication means to connect with each other without having to say? Why are we so interested in the mind and its capacity to create the improbable? What has awakened man's ability to think deeply and use the mind as the main tool for creation? We used to rely on other means for information; today we simply use our mind at a deep conscious level and create our own future. Can we possibly understand communication as having improved with energy in the universe, which is neither visible not touchable? Is it possible that with a mindful and powerful ability to think consciously, something invisible in the universe connects us to bring ideas into reality? Is there a powerful energy that raises men into a higher level of consciousness to think godlike creative thoughts?

Are there frequencies and a powerful energy impermeable in nature? Does this energy cover the entire universe? To understand energy, one must see the entire sphere of the universe as a magnifying energy source in constant vibration, creating energy all around it. This allows the universe to cover and create more energy in a perfect and unified manner, merging with every planet to create a unique form of frequency and vibration with the sum. The sun, as far as we know, is one of the best sources of solar energy in the universe. However, this energy is not the only one. There is an impermeable energy we think we know, which has more powerful energy than the entire solar system. In our logical mind, we ask ourselves: how is it possible that the universe relies on the sun as its main source of energy? Thus, we conclude that there is another energy in this universe whose invisible nature makes it more powerful than what we see coming from the sun. Our mind ponders and ponders many questions about what this energy feels like, looks like or is like. Imagine having something moving and touching you without you being able to see it. Now imagine this energy having the power to create more energy, destroy planets, and even move at the speed of light without being detected? If this is not puzzling enough, I don't know what is. Yet there is an invisible energy in the universe doing this, and we have no idea about what it is. This can only mean that the essence of this energy is invisible by nature and so is everything that has power in this universe. The idea of something of such magnitude will have us wondering until we finally detect its presence in the universe. Then, we will not longer call it dark energy; we could possibly refer to it as bright energy. A hidden force beneath the dimensions of the universe where space is fill with more energy than we know. This energy appears to be dark to us due to its invisible nature but is white with an infinite powerful energy force.

Everything we conceive is in the universe. Creation is considered to be the mother of all invention with a magnifying power to extend and expand itself out into the infinite of more creation. Think of it as an existential power of creation that never stops reproducing itself. Never! We are constantly imagining, creating, having new ideas, and then making those ideas into the world as part of our own reality. This creation is what has made this world possible. Whether it was an idea or simply a form of energy put into action, something was created from something to bring that something into a reality where we live today: Earth, planets, and the universe! I know this is a bit deep for some of you. However, one day, you will understand it all...

When we integrate communication into a new idea for new advancement of the future, we cannot possibly neglect our ancestral contributions, because, since the beginning of time, they have made an impact on our society. Understandably enough, it is with their contribution that we make communication possible today. From the Stone Age to the first signs of language, communication has been intertwined with the past. Although communication has improved with time, there will always be a big gap in communication between the new generation and the baby boomers. How will they feed in? What role will they play in the new technical world, and how will they impact future generations? What role can they play today in the new millennium? when we think of the elderly, many might think of the last days of living here with us until that day comes—death! But that is not always the case. They, too, have made their contribution to society in some form of communication with us, like I have, in the pages of this book, at the age of 65. Therefore, we are not to be excluded from the advancements of technology today.

The future of technology includes all of us, with no exceptions. We should not exclude anyone. Communication must be improved at all levels of education. Education is not just for the technologically savvy or wizards, but for everyone who is driven to learn about any subject of interest. The question is, how do we implement communication and create a better world? Television is a good form of visual communication, and millions watch it every day. Educating the masses only requires intuitive thinking. Another way of teaching the older generation about technology is to simply create a tool that is specifically designed for them. This tool can captivate their imagination and help them learn what is happening in today's world without much complexity. If we are going to create a better world for all, there must be a way to incorporate everyone; no one should be left behind. Old, young, new, and intermediate, all shall be equally included in the future, or we are basically creating a world for those who contribute to society more than others. It could be selfish on our part to distinguish between ages, race, or religion.

Do we firmly believe that AI can create a future, make an impact, and then exclude everyone who is not connected to it with a microchip? If this happens, it could be a plan for extinction. If it is true that knowledge is wisdom, then someone somewhere will create a new generation of savvy computer techs or wizards with ample knowledge to change the world. Then they will exclude those who are not part of the technological world of advancement they have created. How would we approach this if it did happen? What would happen to the rest of humanity that is not part of the conglomerate technical world? Unfortunately, extinction is the possibility.

If ideas can create a brand-new world, why do we need AI to do it for us? What part will AI play in our own imagination of a future we have designed ourselves? This idea will include everyone, homeless, young, old, entrepreneurial, no exceptions. Every existing living human being will fall into the same categorical order, and no one will be left behind. Now, tell me, is there another plan for using AI? Because, if there is, we need to get rid of them and find a way to exclude them from our future. It is my view that AI should make a global impact and nothing else! If we use AI for a global impact, the financial gains will quadruple. Even if we only use AI for technological improvements, its benefits will outweigh all projects in progress today.

Whenever we hear the word "extinction," what comes to mind is our demise at the hands of AI. Either the Earth will disappear, or it will be destroyed by a catastrophic event caused by AI to make it look like it destroyed itself. There is only one way to re-establish a new planet, and that is with a new generation of technologically savvy individuals from around the world who can either change the world or destroy it. They could overpower technology and create their own new world of intelligent, creative individuals. They are the ones who will create AI. If we look at the past, we see that there were some civilizations like the Mayan that disappeared from the planet without any physical explanation. They simply vanished! This could also happen to us. It is possible that for one generation to supersede another, one should end for another to begin. Think about Atlantis or the idea behind creating an AI to handle our weapons or safety. Can we blame AI for destroying a planet or a civilization of people? Crazy or not, it is a thought that should cross our minds carefully.

We can all be in tune with the future or be consciously aware of what is happening around us. However, communication is critical to understanding the nature of machines' operation and intelligence programming with AI. There will always be a way to interface and create better communication with AI. Transferring information is more about mindfulness than it is about intelligence. This ability to use the mind as a form of sending and transmitting information will be the best possible way we could communicate with AI in the future. There is always a way to relay a message with the mind. If two or more are mentally tuned in, when the message is sent, the receiver of the ideas must be ready to receive from the field of nothingness into his mind. They both have to be tuned-in consciously to receive. Perhaps this is how subatomic particles behave during experiments with atoms, electrons, or photons. They are tuned in before the actual separation of the particles begins. To have success with both minds intertwined together, there must also exist mental agility or flexibility that creates alertness or awareness of senses so the brain can communicate information to the mind with some transmission of energy from the neurons to the brain.

Such mental agility can be learned more effectively when in communication with the world of the unknown, the world of magical interpretation, the conscious state of mind with the field of energy. Ideas are how imagination was used to create our world today. Our interpretation of reality was created with ideas. Ideas and innovation are the core of how we created everything we see today. A simple thought became an idea, which then became a physical reality. In the world of interpretation, everything began from nothingness to become something in our presence. This world of nothingness is what we refer to as the emptiness or the vacuum of space.

But is there emptiness in space? Do we not receive ideas and information from the nothingness of space to make it a reality in our world? The truth is, there is no emptiness in space. Everything has an origin. Even the idea of creation came from a thought, which then came from the field of energy that was before that idea was thought of. Everything we know today came from the origin of unknown space-time in the universe. Because the quantum world presents such an enigma to us, we don't fully understand the nature of invisible energy as it should be. Without energy, nothing would exist today, nothing! Everything we do, think, or create has some form of energetic vibration to it, and so do we! With our creation, this energy is expanding with more energy exponentially contributing to the expansion of space time. We are all connected to what is appearing in space today. We too are part of every changes happening within our environment and the cosmos. We, and this energy are one!

There is an energy in space that had an existence before its own. Think of it as a thought, an idea, a creation of something new. Before that new idea or creation came into a thought, its existence was nothing of its own. Where did the idea, creation, or thought come from? It is nearly impossible to think or believe that creation itself has no origin. There must be a form of energy or an invisible thing that makes things happen. The universe was not made from nothing! There was energy, which then created an amplitude of energy to create something new and more powerful than itself. We can call it whatever we want, but that energy must have existed before it burst into a massive force of light!

The primordial essence of life has energy, which is life itself. This life is created from energy or with energy, and that energy cannot be empty itself. There is no emptiness...not in space or even around us. There are energy and information in everything that exists. As a result, information and communication come from what we call the emptiness of space. However, we believe that this space is not empty; it is filled with information. This information is promulgated to us with thoughts, consciousness, and ideas. This can only mean that ideas are a vast, infinitive resource of information in the universe. They are communicated to us from this space-time, or emptiness, as we call it. With a deep understanding of the power from the quantum world of energy, this reveals the true essence of what is possible with our mind. If information is out there in the infinity of space, then what would stop any communication from tapping into our minds? The invisible world has a mystical and powerful energy we have yet to discover in the universe. Like the world of quantum mechanics, where everything is tiny and powerful at the same time, our ideas can fluctuate in and out of space with time. Just as the subatomic particles do when they split but connect no matter where they are, so do ideas come and go in and out of existence until we are ready to grasp them with our minds.

If my reasoning does not deceive me, understanding about nature, life, and the universe can be complex or just simply individual theory. Whatever the case may be, I refuse to believe that life or any aspect of our present existence has no meaning. If it is true that nothingness is part of the universe as we see it, then we are the very essence of that nothingness that has been created with time. Thus, everything created from it will return to be simply that: nothing! With all the new ideas we get to consistently change and create new things, we can say that perhaps the thought of being created from nothingness just doesn't cut it for me. Seeking the truth can only lead us to find it, but are we ever going to be satisfied with it? I sincerely don't think so! The words "seek and you shall find" do not determines the end of our quest. We are curious creatures, and once intelligence has reached such a high plateau in our minds, there is no end to it.

ARTIFICIAL INTELLIGENCE OR HUMANOIDS OF THE FUTURE

WHEN THINKING ABOUT an artificial intelligence and brain interface, think of the perfect humanoid of the future, a human-like creature with our intelligence embedded into its machine brain. Why humanoid? There is no doubt in my mind that we can create a neural link with a machine and brain from the workings of the brain and thus imitate it with a machine. The question many may ask is how? How do we take a machine and make it think like us? Can we design a humanoid with extraordinary capacity, one whose brain capacity will be greater than ours? Can we? The answer is yes! But don't let me scare you. How do we incorporate a human brain with a machine? The answer is simple. First, let's not be hypocritical here. Today we create artificial hearts, we create new organs from stem cells, and we can create as we think, so why is it so difficult to create a human brain with artificial parts to improve thinking and many other mental improvements? If we can build hearing aids, we can enhance the brain to exceed its natural capacity.

Have we not found a new way to select or decide what gender we desire to have? Preferably male. Yet when it comes to persuasion, a female does a better job. Everything that we can think of today has the possibility of becoming a reality. It is mesmerizing to think that in many ways, artificial intelligence is already being used today. The future is already here! AI is in self-driving cars, in robots building cars in factories, in medical equipment, and in many other sectors. With stem cells, we can heal, regenerate new organs, restore damaged cells, and even build new organs. The future is promising. These are all done with a computer system designed especially for medical purposes. The future of medicine will be enhanced with the use of AI to heal the body, improve our minds, and create a better world for all of us.

However, there will be some consequences while improving the way we use medicine in the health industry. How about our spiritual well-being and growth; when will AI become compassionate, caring, or loving like a human being? Are we, the human species, becoming less compassionate because our emotions are already linked to a machine that experiences nothing emotionally? Have we become desensitized? If we continue to act like mechanical humanoid robots with our tech toys, we will wake up one day to find that our human touch and ability to have compassion have disappeared. Sad as it may sound, we are on the verge of a deep emotional transformation with the use of AI. The only thing that can save us from losing our human touch will be to continue to be aware of our mental, emotional, and physical state, thus becoming more consciously aware of our behavior. In the end, consciousness will be the only cognitive transformation keeping us from becoming machine-like human robots.

Today we are experiencing a big change with our children's behavior as they get more in tune with their electronic toys, computers, games, and all the programs available online. The transformation in their behavior has already begun. However, it is up to us to understand what is happening, what is inappropriate, and what needs to be done to stop the fever of electronic devices from overtaking our brain and our ability to function like normal, caring human beings.

Every single aspect of our lives today is either monitored, controlled, or being watched by a new computerized device invented with us in mind. We can prudently use and enhance our intelligence and that of AI. However, the use of intelligence from a machine can also have devastating consequences for us, both mentally and spiritually. We can lose complete touch of our well-being as we think and act more like an AI or human robot. Many experiments of the mind have shown a change in the perception and beliefs of people about themselves and the environment! Nevertheless, if we are not well educated about the psychological aspects of the use of AI, it may have a negative impact or irreversible negative effect on the psychological well-being of the mind.

Neither intelligence nor knowledge guarantees perfection in trials of the mind and machine intelligence. Until we know more about the outcomes of AI's use, our responsibility is to take precautionary measures. At some point, we must learn to distinguish between us, and a machine operating like us. In the end, it will not be AI who overpowers us, but rather, it will be us giving AI the power to control and manipulate our lives and our limitations. If something goes wrong, the machine is not to be blamed. We are responsible for each choice and decision we make for our future. It will not be wise to give control of our lives to a machine that then destroys us when it exceeds our intelligence. If we are worried about human extinction by machine, we need to realize that we are the only ones responsible for such an outcome. Machines can only do what we teach them to do. All else is man's responsibility, not the machine. We create everything that either brings progress or destruction to us. In the final decision, we have a say.

Therefore, any outcome is also created by us, our intelligence, our programming, our ideas, our creation, and our planning. Man creates, and man destroys. Without us, there is no AI; there is no future, no intelligent machine, no interface, no human brain or microchip implant. Every idea was first thought and then created and implemented. Nothing ever stands on its own. We imagine, we create, we invent, and we make things happen.

We are making AI a part of our everyday lives. We are the co-creators of an intelligent machine that can one day operate and think like us. These superintelligent machines are not too far from dominating us soon. It is our human curiosity driving us to learn more about ourselves and our future that makes us push the limitations we once thought we had. Our present evolution and advancements today can be a perfect testimony to our eager needs for discovering the unknown.

Perhaps the deep need to know more about ETs has led us to create a mechanical simulation of what we think ETs are like. Our eager desire to meet our opposites has driven us to make it happen with our own imagination, creativity, and intelligence connected to machines. Movies aren't enough, so we have created ETs to make them more real. Could it be that our fascination with the unknown drives us to create with our minds what we think is or is not real about the ETs? Or do we lack understanding of ourselves? Could AI be a perfect interpretation of ETs? Or are the ETs and AI a transformation of us with higher intelligence and higher learning capacity? Another possibility is that we have discovered what ETs are, and now our intelligence has caught up with theirs as we replicate what we know or have learned from them. In all probabilities, the odds are in our favor. If we continue to reinvent ourselves in the images of others, there is no danger for us, the human race; we are all safe!

ARE WE FULLY CONSCIOUS HUMAN BEINGS, OR ARE WE NO DIFFERENT THAN MACHINES TODAY?

AS WE ATTEMPT to create the future, we are not using our highest level of consciousness to deal with the NEW INVENTIONs consisting of man and machine – AI. Presently, we are more in tune with making a machine think and act like us than learning what makes us think. The big picture is not yet clear. We must ask ourselves if the consequences of connecting our brain to a machine could be more emotionally detrimental than beneficial. Then we can use logic to prepare ourselves for the outcome. In the process, if we lose our human touch, unconsciously, we could become like them, simply human robots thinking and acting as commanded by them—machines. If we interface our brain to a machine, there would be no reason not to think, respond, and communicate in the same way it does with us. Whether it is a machine or a device, it makes no difference.

Are some of these devices already changing the way we think and act amongst each other? Are we simply imitating machines as we use them? As eccentric as we are in this modern world, there are those aspects of our new technological advancements that question our integrity and human touch. But we worry not about human integrity today; thus, our ultimate dream is to exceed intelligence and speed-of-light communication. But such is the process of evolution; we are always running ahead of the next game.

How fast are we making advancements? Are the ETs among us? Have we finally established communication with them and, as a result, making great improvements in our present evolutionary civilization? Has consciousness finally showed man his ability to tap into a level of his mind where he alone can make and discover miracles with his senses and the use of his brain? The fact that evolution has taken a greater step towards advancement with us does not justify our advanced intelligence; something else has happened, and as a result, we have become more aware of our mental powers and the capacity of our minds to create the unthinkable. But how did this transformation take place during our lifetime? When or why didn't we know this a century ago? Why now? Deep thinking allows us to discover the truth, and the truth is never hidden; it just hasn't been noticed before. I say there is more than what we know today, and reality is not what we think it is; therefore, we are the true catalysts of our own progress with a little help from our friends, the ETs.

Establishing communication with a device or machine is not the same as human interaction or communication. A machine can only interact and communicate with a human brain if there is a quantum connection at a conscious level of the mind. Although it may sound like a crazy idea, communication with a device must be at a telepathic level to create a good rapport. After all, a machine doesn't talk. AI must mimic us, and we then learn from AI. Only in such a way can we achieve some form of communication.

Quantum communication between the brain and

a machine happens when the two have made a perfect connection, correlating thoughts at the same speed and at the same time. This will help the brain and the machine both enhance communication with practice. It is my personal belief that what the future of communication with AI holds for us will take us by surprise. We use the mind to think and relate information to others, and nothing says that a similar process cannot be administered with a machine. Must this machine be conscious to understand and relate back to us? Could it simply use interpretation by learning, or would it be aware of what we are thinking at some point during this communication? So much needs to be known or understood about how communication with a machine can be carried out successfully. Currently, such a miraculous act of response seems nearly impossible. What have we yet to learn?

To have any successful communication between us, human beings, all we require is a mutual understanding of the fact or conversation. But first, we have to recognize and accept the communication as equal and have similar knowledge of the idea or thought being discussed. Then there is an emotional or intellectual response. Consequently, consciousness plays a big role in communication of any nature. But how would it affect our communication with a machine or interfacing with it?

Scientists refer to this as the perfect communication between electrons. During this experimental trial, there are no observers present. Quantum mechanics seems to have a very private sequence of interactions. If any observation from outside happens, there is an immediate change in reaction, information, or even performance between the electrons. Every aspect of our life is a quantum reaction. There isn't one without the other. Every thought we have is quantum. This is because thoughts generate energy through the neuropathway; this energy then connects to more neurons, and the energy in the brain multiplies and produces a higher level of thought. These thoughts then spread like waves because thoughts have energy, and the frequency of that energy spreads to many angles of the sphere from where it was first created. Think of it as thought patterns. They form, and then they continue to spread until there is a new generation of similar thoughts. Would AI use thoughts in a similar manner? If not, how will it assimilate thoughts in its program? If AI can learn and improve on its own, there is a probability that it has acquired some form of conscious understanding of the thoughts it is creating with us.

Although we have created this image of AI as machine, it is not about interaction with a device; it is more about the mind connecting with things outside of itself. We don't fully comprehend its nature with our own logical mind. Whether it is a machine or the mind, there could be no interaction without the field that connects them together. There is more to our conscious mind than what we currently understand about it.

Perhaps, this is where we have it wrong, thinking our attachments are an emotional response created by the mind. If we use AI, once we have become more responsive, attachment will no longer be a problem. At a conscious level of the mind, we will be able to disconnect from those emotions with AI in our brain. Because of our mind interaction with machines, restoring our emotional well-being and behavior will help us learn how to detach and use communication to interpret what is happening to us at a very deep emotional level. Such a behavioral response would first have to be learned by the machine or AI, which would then reinforce it in us. As we learn to detach from our old beliefs, our ability to communicate not only with machines but between ourselves will get better. We will have a completely new understanding of the power of using our mind to control all emotions, our intelligence, plus all the energy coming to us as a result of been mentally connected to a machine.

If the brain is like a software, and it is wired to a machine, its performance, enhancement, thinking patterns, and reactions will also have to change. Not only will AI learn from us, but we will also learn how to adjust our ways to be able to communicate with a machine-AI.

I WILL NOW INVITE YOU
TO PONDER THIS...

THE YEAR NOW AND TODAY is 2019.

Since 1947, AI has been the topic of conversation for philosophers and scientists alike. That year, Alan Turing questioned the idea of whether it was possible to construct or simulate the human mind. Today, that question is still unanswered. However, with the mystery of Areal 51 and all the unanswered questions about aliens, one can't help but wonder if we have been in contact with intelligent life other than our own. Machine intelligence is no different than human intelligence because, we, the human species, are the ones programming it. But what about possible contact with other species? Have we exchanged valuable data with aliens? The answer to this question is about to surprise you.

Imagine that it is the year 1970. A baby boy is born somewhere in Europe. He is of Arabic and Italian descent. His parents have him out of wedlock and were not planning on having a baby, or let's say his father is a king in the Arab world, but his mother is a young Italian woman who got pregnant from this king. After the birth, at the hospital, one of the nurses is asked to exchange the baby for a newborn. The newborn is part of an experiment going on behind the scenes with the Italian government, with international cooperation, for a project called "The Hybrid"!

Thus, a new hybrid is integrated into our society: a seemingly normal baby whose father is part alien. This child is trained, educated, and treated as one of ours; however, there is a slight chance that his teachers may see him as being different from the rest, or he might have difficulties with the other pupils. He has some special characteristics or is smarter than the other children.

When the child is grown and becomes an adult, beneath the surface, it appears that he has special gifts. He can do things that others cannot do. He gets away with violations of the law and appears to have supernatural powers. He is also very sexually in tune. In other words, he comingles with human females, breeding with them and creating more of his kind. The only difference is that females are driven to him, but he is completely detached from any feelings of commitment to them.

Many years later, now a 49-year-old male, this hybrid has accomplished the task of multiplying his own hybrid with ours, and we see what appears to be children with very different behavior and characteristics than ours. The only problem we have is that we cannot tell the difference between such hybrids and normal children. Soon, the population is large. They look different than our children, but we blame it on the new generation or food.

They are now connected to a new network of hybrids who live among us for more than 50 years. They can't be identified because, unlike us, they have the special power to hide and become invisible in moments of crisis or confrontations with the law. The truth is that they are being protected by the same group of international agencies involves in this program: "The Hybrid"!

If you noticed, every year, thousands of children go missing. They are not missing; they are simply put into a new program that protects them and keeps them away from their parents while they are trained in a new method of mental transformation for our future society. This may sound like a scene from a movie, but what if this is happening today? Do you think this is sex trafficking or prostitution? No…we are talking about a new trend of mind transformation coming from a very different circle of control and mentality. What if these hybrids are already here? They are among us, and they are training our children to become insensitive, detached, without emotions, and able or willing to do whatever they are told to do without reservation.

Consequently, the group of adults involved in this program could use it to manipulate the system and hide what they are doing. Because of their power, they could not be prosecuted for any of the damaging behavior used while training the children of the new generation. They are training these children to respond to them psychologically, mentally, and sexually. Their purpose is to continue improving the hybrid kids so they can transform the world, and sexual behaviors are taught at a very early age to seduce them into having more sex and increase the population of the next generation of hybrids.

But this is where AI will make a difference in this world. Trained to be a subagent, AI can detect who is involved, what is happening, where it is happening, and what kind of training is taking place. The new world of AI subagents is useful for all purposes, including areas where privacy is extremely tight. Nothing will ever bypass the intelligence of a machine whose job is to predict the behavior and nature of humans as well as hybrids here on Earth.

Although this type of program can be kept under highly secure privacy, there is no doubt that we can deal with any case of hybrids and humans cooperating. With AI, we can use new tactics to help penetrate the system and gather as much information as possible before entering dangerous territory.

How do we ameliorate the idea of an AI operative working as a subagent with a human agent? If we can manage to incorporate the brain with a machine successfully; we may have a change at creating the perfect partner for an agent. The idea of a subagent could be more efficient when facing international political or violations of any kind. Using AI as a subagent will prevent any fetal death or endanger to the human agent, thus it can help us establish a counterintelligence cooperative operation with more success. During international or domestic threat, we have used counterintelligence to keep an eye on suspects or the authenticity of intelligence agents who were tempted to sale information to the opposite side. We can now prevent this with an AI subagent who instead acts as an informant to relate back all pertaining information to the agent in charge of the case.

In such a case, who will be the best agent assign for this case? A man, or a woman? Because of the nature of international threat or terrorism today, we can say that a man would be the best choice, however, considering how a super intelligent subagent will know what to do; it will be feasible to use a woman. This woman will have to be a highly intelligent subagent with great sex-appeal to attract the opposite and leave no room for suspicions. This intelligence tracking can be done with a subagent with a microchip already implanted but has not previous knowledge of the case. She could be trained to know only the details of the agenda to be carry on. Nothing else. It is possible for an AI to be more efficient in the field of investigation; there will be no suspicion, less risk or little possibility of getting in and out without been harmed. For the human agent it will be highly probable to encounter more complications. There is a reason why affluent man and diplomats are more attracted to women of beauty. However, there is a high price to be pay for these beauties, and it isn't always money. Information exchange has a high value, and there are those who are willing to pay for it at a very high price.

IN CONCLUSION, I WOULD LIKE TO SAY...

WHAT IF ONE of your children were part human and part hybrid? How would you know if they were wired with a tiny microchip inside their brain to make them smarter at school, without your permission? How would you know what would make them act and think different? How could you tell? Would it be their behavior, their intelligence, their actions, or would they simply be recognizable by sight?

The truth is that an AI can identify any case like this. This is the case of a machine versus human intelligence. Which one will you trust? If we are referring only to hybrids, you will have to trust AI. Humans can hide the truth. AI, on the other hand, is not capable of breaking the rules or disrupting the course of its programming at any time without proper commands unless it gets smarter than us. The ultimate question of the day among us is: will AI one day change and completely transform our future in ways that will forever change how we see ourselves? Evolution will one day take precedence as we wonder what a true human being is.

Have no fear; the future is ours, and AI is here to change, correct, create, enhance, or even multiply itself to change the way we live.

Written by Frances Mahan
From April of 2017 to…
This day of July 31-2019